WHAT'S
GOTTEN
INTO US?

WHAT'S GOTTEN INTO US?

Staying Healthy in a Toxic World

McKay Jenkins

RANDOM HOUSE
NEW YORK

Published in the United States by Random House,
an imprint of The Random House Publishing Group,
a division of Random House, Inc., New York.

RANDOM HOUSE and colophon are registered
trademarks of Random House, Inc.

LIBRARY OF CONGRESS CATALOGING-IN-PUBLICATION DATA
Jenkins, McKay
What's gotten into us?: staying healthy in a
toxic world / McKay Jenkins.
p. cm.
Includes bibliographical references and index.
ISBN 978-1-4000-6803-6
eBook ISBN 978-0-679-60497-6
1. Environmental health—United States.
2. Environmental toxicology—United States.
I. Title.
RA566.3.J46 2011   362.196'98—dc22   2010026778

Printed in the United States of America
on acid-free paper

www.atrandom.com

9 8 7 6 5 4 3 2 1

FIRST EDITION

Book design by Mary A. Wirth

FOR KATHERINE, STEEDMAN, AND ANNALISA
WITH LOVE, DEVOTION, AND GRATITUDE

Contents

WHAT'S GOTTEN INTO US?

Prologue

On a crisp fall afternoon a couple of years ago, I went in for a routine two-year checkup with my internist. Everything seemed to be fine: My home life was happy and nurturing. I had never smoked. I ate right, got plenty of rest, and had been a dedicated runner and cyclist my entire adult life. Save for the usual aches and pains, nothing had ever been wrong with my body, and as long as I was smart about it, I figured, I'd still be riding my Fausto Coppi racing bike well into my eighties.

My only complaint, I told my doctor, was a faint tightness in my hip that I had felt off and on for two years—and odd, sharp twinges between my left thigh, knee, and shin that occasionally accompanied it. Sometimes the skin on my leg itched. Sometimes it burned. Sometimes the ligaments in my knee hurt. I'd consulted with a dermatologist months before, but had gotten no answers. Were these symptoms related? My internist was perplexed. Perhaps it was an "overuse injury," he said, something I'd developed from too much running in the woods or riding the rural roads near my home. Like anyone who has tried to stay in shape through their postcollege years, I was familiar with such aches and pains. As the years went by, fewer and fewer were the days when I *didn't* feel some minor muscle or joint ache after even a light workout. I was getting older, and so was the machinery.

My internist looked me over and agreed that my pains were probably related to exercise, and he suggested I see an orthopedist at a nearby sports medicine clinic. I called and got an appointment that very morning. I walked into the orthopod's office ready for a quick diagnosis and a pat on the head. "Someone as fit as you can expect to have occasional ligament stress," I expected him to say. "Here's the name of a physical therapist. Go get a massage, and check in with me on your seventy-fifth birthday."

This is not what he said.

After hearing my description of the pain, the orthopedist rotated my hip and knee a couple of times. He seemed puzzled. This didn't seem like a joint problem, and in any case the pain in my knee was probably "referred" pain radiating from my hip. He suggested I get an MRI to help him see a bit more clearly what was going on with my soft tissue.

Okay, I thought: people get MRIs all the time. Especially athletes. They'd probably just find a slight tear in some connective tissue, I'd buy a new pair of running shoes, and away I'd go. Worst case? A little minor surgery to fix an abraded tendon. I got an appointment that afternoon, spent forty-five unpleasant minutes inside a clanging metal tube, and went home to wait for the results—which, given the routine nature of the exam, I figured would take a few days, or even weeks.

I was standing in my living room when the phone rang just a few hours later. This was an awfully quick turnaround, I thought, looking at the caller ID. These lab techs must be having a pretty light day at the office. But when I picked up the phone and heard the orthopedist's voice, I knew even before he spoke that something was amiss.

Hello, Mr. Jenkins, he said, then paused. You have a suspicious mass in your abdomen, he said. It's growing inside your left hip. Here is the number for an oncologist. You need to call him right away.

What can you say about such moments? I remember hanging up the phone. I remember looking at my wife, Katherine, and looking at my children putting together a puzzle on the floor in the next room. My son was four, my daughter not yet eighteen months. I fell apart.

Far worse than my fears for myself were my worries about my kids. How would they get by without a father? Trying to protect them from the initial shock of the news, Katherine and I took turns taking our cell phones outside to talk to doctors and loved ones. Standing in the yard, trying to set up a date for a CT scan, I would look through the living room window to see my kids playing together. I imagined the same scene, with me gone. I felt like a ghost.

Katherine and I passed the next three weeks in a kind of silent panic. We spent anxious hours on the Internet, blindly researching what this thing was that was growing between my hip and my belly. This was a very bad idea. We called every doctor we knew.

I held it together enough to keep teaching my classes at the University of Delaware, which is about an hour north of our home in Baltimore. One day, when I returned to my office after a morning class, there was a message on my phone. It was from Katherine. I found an oncologist, she said, but you need to get here fast; he's really busy and has only one opening this week. I ran into my classroom, scrawled "Class canceled" on the blackboard, and dashed off to my car.

When I got to the hospital, Katherine was already there, waiting in the parking lot. We hugged, and went inside. I was short of breath. A nurse escorted us into an examination room, and a few minutes later, the oncologist strode in with my MRI images under his arm. He slapped them up on a backlit screen. You see this? he asked. That's your hip. Now, see that round thing next to it? That's not supposed to be there.

He reached over to the examination table and began scribbling diagrams on a sheet of sanitary paper. He sketched out what he thought was going on. Although he couldn't be certain without further tests, the tumor was likely a soft-tissue sarcoma, an ugly cancer of the fibers connecting my hip to the muscles and nerve tissues of my left leg. The prognosis depended on how big the tumor was, where precisely it was growing, exactly how aggressive it was. I could get a biopsy of the tumor, which would provide a look at its cellular profile, but there was always the risk of a biopsy needle inadvertently spreading malignant cells outside the tumor itself. What to do?

At worst, this was, well, very bad. At best, a surgeon could cut out the tumor, but might be compelled to sever my femoral nerve, the trunk line that connects the nerves in the leg to the spine. Which meant I would probably never run or ride my bike again. And then I'd have to remain vigilant to see if the cancer returned. The doctor ripped off the sheet of sanitary paper, now four feet long, and handed it to me.

As he continued to speak, I felt myself leaving my body. My mouth continued to move, but the rest of me was floating, looking in from the outside. I saw myself shake hands with the doctor. I saw Katherine take my hand and lead me outside. It wasn't until we made it back to the parking lot and I dialed my brother on my cell phone that I returned to my body. "Brian, it's McKay," I said. "I just came from a doctor's office. I have cancer." The floodgates opened.

To this day, the ride home has remained indelible. Once-generic landmarks still vibrate with the terror I felt that day: the gas station where I stopped to call my friend Wes. The curb by our home where I called my friend Tom. Rather than relieving my anxiety, each telling made the story more concrete. This was really happening. But how? This was not a grinding descent into illness; it was a bolt from the blue. I did not feel sick, and never had. My mind raced. How could I possibly have cancer?

Beyond the panic, of course, was a question. Where had this thing, this "suspicious mass," come from? No one seemed to know. Not my primary care doctor, not my orthopedist, not even my on-cologist. In medicine, cause and effect are not always clear.

After weeks of scrambling and using every connection I could muster, I found a slot on the schedule of a renowned surgeon at a New York medical center. Katherine and I drove north, dropped our kids off at their aunt and uncle's apartment, and went to the hospital.

In the morning hours before my operation, I sat on a couch in a waiting room in a light-blue surgical gown. I wore headphones, listening to the Dalai Lama offering counsel about facing one's own death. Life is a series of transitions, he said. Dying is one of them. It's vitally important to remain clear-eyed during these times, to see things Just As They Are.

Okay, I thought. I will do my best. But seeing things clearly is not always so easy. Especially when it comes to understanding illness and its root causes.

At some point, I looked up to see the outstretched right palms of a pair of researchers, a man and a woman in their twenties, clipboards at their sides and kind, self-conscious expressions on their faces. I took off my headphones and we shook hands. We'd like to ask you a few questions before you go into surgery, they said. Sure, I said. I was feeling strangely serene and, I confess, a bit melodramatic: If I'm going to die, I thought, the least I can do is be a model for others. Go out gracefully. Make the Dalai Lama proud.

The researchers sat down on a couch across from Katherine and me, placed their clipboards on their laps, and began probing. The first questions were pretty standard: What ethnic group best describes you? Um, white. How far did you make it in school? I have a PhD, I said. How many packs of cigarettes have you smoked per day, on average? None, I said. Ever. The researchers nodded, and scribbled. How much alcohol? A couple of beers a week, I said. I managed a wan smile. Please don't tell me I have a tumor because I drink beer.

Then the questions changed, from ones I had been asked by doctors dozens of times before to ones I had never been asked in my life.

How much exposure had I had to toxic chemicals and other

contaminants? In my life? I asked. This seemed like an odd question. What kind of chemicals do you mean? The researcher began reading from a list, which turned out to be long. Some things I had heard of, many others I had not. Metal filings? Asbestos dust? Cutting oils? I didn't think so. What's a cutting oil? How about gasoline exhaust? Asphalt? Foam insulation? Natural gas fumes?

Where was this going?

I was not a machinist, or a car mechanic, or a building contractor. The words kept coming. Vinyl chloride? I wasn't sure. What was that? How about plastics? Are you kidding? Everything is made of plastic. Dry-cleaning agents? I shot a glance at Katherine and managed a smile; I was not exactly known for my natty attire, and I hadn't darkened a dry cleaner's door in years. On and on it went. Detergents or fumes from plastic meat wrap? Benzene or other solvents? Formaldehyde? Varnishes? Adhesives? Lacquers? Glues? Acrylic or oil paints? Inks or dyes? Tanning solutions? Cotton textiles? Fiberglass? Bug killers or pesticides? Weed killers or herbicides? Heat-transfer fluids? Hydraulic lubricants? Electric fluids? Flame retardants?

By now I had begun to feel distinctly uncomfortable. Not about my history of "industrial" exposures, which were nonexistent, but about the myriad, and mostly invisible, chemicals the researchers seemed to be curious about. What was a flame retardant, exactly, and how in the world would I know if I had been exposed to one? I had never used pesticides, but Lord knows there were plenty in my neighborhood.

The questions shifted again.

We'd like to ask you about your job history, they said. What had I done before I became a professor? Well, I thought, I hadn't exactly been working in a chlorine plant. Before teaching English, I spent a decade writing for newspapers. I'd spent a lot of college summers waiting tables in restaurants and, before that, washing dishes. Oh, wait—there was that one summer after college when I worked in a garage, pumping gas and changing people's oil. Could that have been it? And was that the point of all these questions? To find a single cause of this tumor? If so, why would they be so interested in what I had done for work thirty years before?

Again, the researchers changed their tack. Now we'd like to ask you about the places you've lived, they said. If my employment history seemed fairly benign, I had spent a lot of time living in big, industrial cities. Again, going backward in time: Baltimore, Philadelphia, central New Jersey, Atlanta, Seattle, Annapolis, Manhattan, western Massachusetts, suburban New York.

Any cancer in my family history? My paternal grandfather had died of prostate cancer at an advanced age, I said. My mother had had malignant melanomas removed from her skin. I had, thus far, showed signs of neither.

Had I ever lived within a mile of any kind of waste incinerator, either at a city dump or a hospital? Not that I knew of. I had never worked in an industrial plant; wasn't that where all these chemicals were concentrated? A decade earlier, I had built a couple of canoes, but I had done all my fiberglassing outdoors. I had painted a bedroom or two. Who hadn't? Yet the longer the questioning went on, the more I began to realize that I didn't have the faintest idea about how many of these chemicals I had come into contact with over the years. After all, just because I had never worked in a factory did not mean that I hadn't been exposed to the products these factories made. But once chemicals were turned into products, they stayed put. Didn't they?

The researchers, to be fair, were not there to talk about these larger questions. They were there to ask their questions, and when they were finished, they stood up to leave. The young guy gave me a parting look that seemed slightly melancholy, and they said good-bye.

A couple of hours later, a doctor led me into the operating room, and I lay down on the table. How in God's name did I end up here, I wondered. The room was remarkably cold. Around me, a dozen faces in blue hats and masks scurried around, tending to monitors and swinging trays of instruments at my bedside. How you doin'?

asked a man who introduced himself as my anesthesiologist. Fine, I said, but I could use a beer right about now.

I'll give you something better than that, he said, hooking me up to an IV. Do me a favor and count backward from twenty, would you?

I made it to nineteen.

A moment later, it seemed, I awoke. My eyes felt fuzzy, and blurred by bright overhead lights. Where was I? I blinked. There, at the foot of my bed, stood Katherine and my surgeon. Both were beaming. Something must have gone well, I thought. You're a lucky man, the doctor said. The tumor was as big as an orange, but it turned out to be growing out of a nerve cell rather than a muscle cell. We sent a slice of it down to the lab; it turned out to be benign. Of a hundred cases like this, about four turn out this way. Not only that, we managed to peel the tumor off your femoral nerve. Once you recover, you can get back to running and riding your bike. You're a very lucky man.

And so I was.

As joyful as my outcome had been, I was left feeling somehow bereft. Had this whole thing been bad luck? Where had this tumor, this navel orange, come from? It wasn't until I'd answered the hospital questionnaire that I had ever even considered the vast arrays of chemicals I had been exposed to over the years. Was it possible that the questions constituted a trail of bread crumbs that could lead me to some answers? Suddenly, these questions began to take on a whole new sense of urgency.

From the moment I thought I had cancer to the moment I learned I didn't—an extremely long month, as it turned out—I felt like I had been initiated into a vast secret society, one whose members had been stamped with a disease whose origins were utterly and maddeningly opaque—to themselves, and frequently even to their doctors. Colleagues, friends, students, family members I hadn't heard from in years—everyone, it seemed, had had a brush with cancer or was intimately connected to someone who had. In the four years that have passed since my surgery, cancer has burned its way through a swath of my family and friends. My beloved aunt Julie recently passed from a combination of breast cancer, bone marrow cancer, and Parkinson's disease. My friend Scott, still in his forties, just learned that he has pancreatic cancer. Like me, he has small children. So do Leah and Suzie and Susan, young women who have both recently suffered terrible bouts with breast cancer. My cousin's husband, Phil, died of a brain tumor before his fortieth birthday. He left a wife and a young daughter. And on and on and on.

What is going on here?

I used to think that serious illness was something rare, something that touched only the unlucky, those who drew the short stick. No longer. Now it had begun to seem like the world was filled with two kinds of people: those who had received their diagnosis and those who had yet to receive it. My brother Denny, a surgeon, has a dry expression for this: We're all pre-op. Every one of us.

No one goes through a cancer scare without experiencing a kind of awakening. Here's what mine looked like: I went from being a passive observer of other people's suffering to feeling an intimate desire to prevent that suffering. I wanted to know if there were root causes. I wanted to try to see things just as they are, how they came to be that way, and what I could do to protect myself and my children. This book is the result.

To begin with, it's worth thinking about what a relatively short time we've been swimming in synthetic chemicals. In 1992, seven years after the *Titanic* was discovered by a French-American team off the coast of Newfoundland, the French government sought out the few survivors and their heirs to try to return some of the bounty—pince-nez spectacles, hairpins, ivory combs—that divers had discovered on the sunken ship. But what caught my eye about this story was a comment made by Charles Josselin, secretary of the French merchant marine.

"What most struck me," Mr. Josselin said, "is that in such a long list of day-to-day objects, there was nothing made of plastic."

Not a single piece of plastic on the grandest luxury liner of the day. And this was less than ninety years ago.

The Synthetic Century, let us say, has been full of grand achievements and equally grand consequences, many of them unintended. In 1918, a scientist named Fritz Haber won the Nobel Prize for figuring out how to make synthetic nitrogen, a key component of soil, and thus "improving the standards of agriculture and the well-being of mankind." But during World War I, his technology also helped Germany make bombs from synthetic nitrate and, later, poison chlorine and phosgene gas. In World War II, Hitler used another one of Haber's compounds, Zyklon B, in Nazi concentration camps. After the wars, synthetic fertilizers paved the way for the explosion of industrial-scale agribusiness, which has, in turn, created great wealth but also unprecedented levels of pollution, monoculture, and processed foods.

In his book *The Omnivore's Dilemma*, Michael Pollan outlines the way our industrial food chain floats on an ocean of cheap oil. This is also true of our vast array of consumer products. Although coal companies in the mid-1800s were processing coal gas for lighting and synthesizing other products like dyes, this was but a baby step compared to what happened a hundred years later. Since World War II, Big Oil and, more recently, Big Coal and Big Natural Gas, have supplied our economy not just with energy for our homes and cars but with the very building blocks of our domestic lives: not only our plastics but our fertilizers and pesticides, our furniture, our personal care products, even our clothing. Consider this: in the last twenty-five years, the country's consumption of synthetic chemicals has increased 8,200 percent.

By the end of the Second World War, everything, it seemed, was being made from petrochemicals: fertilizers and pesticides for the farm and garden; plastics for everything from shrink wrap to baby teethers to car dashboards; treatments for fabrics. "The synthetics are changing life imperceptibly—changing it, in fact, as profoundly as it was ever changed by the machine," *The New York*

Times reported in the spring of 1940. "The gaudy fountain pens and cigarette holders, the lacquers, the well-designed and attractive costume jewelry in the five-and-ten stores, the curtains at the living room's windows, the steering wheels that we clutch in our cars, the drugs that cure us from some of the more deadly diseases, the dyes that outdo the rainbow, the camphor that once came from the trees, the musk that once cost $700 a pound, the 'glass' in the bathroom—all are synthetics. To the man in the street, 'synthetic' still means a tricky substitute for something authentic—an unfortunate relic of prohibition when raw and fiery alcohol was inexpertly flavored to produce a potent but unconvincing gin. It is time that we gave synthesis its true meaning—a putting together, a control over matter so perfect that men are no longer utterly dependent on animals, plants and the crust of the earth for food, raiment and structural material."

The trouble with such rapid proliferation of products made from petrochemicals, of course, has been that the production and use of synthetic chemicals has vastly outpaced our ability to monitor their effects on our health and the environment. We learned to love what chemicals could make; we just never bothered to wonder if there could be a downside. By the mid-1970s, there were some 62,000 chemicals in use; today the number is thought to be closer to 80,000. The EPA has a full set of toxicity information for just 7 percent of these chemicals, and the U.S. chemical industry, a $637-billion-a-year business, is so woefully underregulated that 99 percent of chemicals in use today have never been tested for their effects on human health. Fewer than 3 percent of these chemicals have ever been tested for carcinogenicity. Far fewer (or none) have been assessed for their effect on things like the human endocrine system or reproductive health. In recent years, as we shall see, the news from these fronts has not been promising.

And since these chemicals—and the products they are turned into—are designed not to break down, they have a way of showing up in some strange places. In 1997, Charles Moore, an oceanographer with the Algalita Marine Research Foundation, found himself navigating a patch of the Pacific Ocean so full of garbage that pieces of plastic—bottle caps, bits of shredded bottles—outnumbered plankton by six to one. Much of this garbage is in the form of "nurdles"—little plastic pellets used as the building material in all kinds of consumer goods. Some 5.5 quadrillion—I've never written a number that large before—of these nurdles are manufactured every year, and lots of them end up in the ocean. Moore has since estimated the amount of debris that has washed off the coasts of Asia and the western United States to be something on the order of 100 million tons. The Great Pacific Garbage Patch, he says, is now the size of Africa, and has become "a toilet bowl that never flushes."

I was born in 1963, the year after Rachel Carson published her landmark book *Silent Spring*, first alerting the world to its saturation in industrial chemicals. A biologist and essayist, Carson set out in *Silent Spring* to chronicle just how widespread, and how dangerous, synthetic chemicals had become. Man had learned enough about chemicals to create enormously potent products, she warned, but had failed to imagine the consequences of their widespread use. "Can anyone believe it is possible to lay down such a barrage of poisons on the surface of the earth without making it unfit for life?" she asked.

Carson's stories, gleaned from newspaper accounts and scientific journals, were ominous, and told with unblinking directness: In Florida, two children find an empty bag of the insecticide parathion and use it to repair a swing. Both die soon afterward. A chemist checking the toxicity of parathion tests it on himself by swallowing less than a hundredth of an ounce. Paralysis sets in so quickly he doesn't have time to reach for an antidote, and he dies. At the time Carson wrote, 7 million pounds of parathion were being applied every year in the United States; the amount used in California alone, one scientist said, was enough to provide a lethal dose for ten times the world's population.

Carson's book was savaged by the chemical industry, which reached for rhetoric as dramatic as Carson's own. *Silent Spring*'s claims about the dangers of pesticides missed the greater threat, the industry claimed, of "hordes of insects that can denude our forests, sweep over our crop lands, ravage our food supply and leave in their wake a train of destitution and hunger, conveying to an undernourished population the major disease scourges of mankind."

In addition to pesticides, the decade before I was born had ushered in the widespread use of polychlorinated biphenyls, or PCBs, the oily fluids used in everything from electrical transformers to window caulk. At that time, researchers tested PCBs for toxicity by dumping them into a bucket of fish. If, after five days, more than half the fish were still alive, the chemicals were considered safe.

In short, the middle decades of the twentieth century were a kind of golden age of synthetic chemicals, a time when DuPont's "better living through chemistry" was considered the coin of the realm. I grew up in Yonkers, New York, where a Sherwin-Williams paint store had a neon sign of a giant can pouring red paint over a blue globe. The sign read: "Cover the Earth." And so we have. By 1940, American companies were already making 300 million pounds of plastic. We wrapped our food in cellophane. We wrapped our legs in nylon. We gave ourselves over to vinyl. Styrofoam. Plexiglas. Polyethylene. Polypropylene. Polyurethane. By 1960, plastic production was up to 6 billion pounds. In 1928, a man named Waldo Semon invented polyvinyl chloride, or PVC. When he died, at the age of one hundred, in 1999, worldwide production of PVC, which is used to make everything from clothing to car interiors to children's toys, had reached 44 billion tons a year.

By 2004, the U.S. chemical industry was producing more than 138 billion pounds of seven petrochemicals—ethylene, propylene, butylenes, benzene, toluene, xylenes, and methane—that form the building blocks for tens of thousands of consumer products, from gasoline to rubber to lipstick. Today, industries worldwide generate 300 billion pounds of plastics a year. You've probably come across numbers like these: Americans throw away 100 billion plastic bags a year—the equivalent of dumping nearly 12 million barrels of oil. We throw away 50 billion plastic water bottles a year.

In 1990, the National Academy of Sciences estimated that in the United States alone, toxic flame retardants were added to 600 million square yards of upholstery fabrics each year—enough to cover Washington, D.C., three layers deep. And that's just for sofas and car seats. Every year, some 40 million pounds of the herbicide 2,4-D—a primary constituent of the infamous Agent Orange once used to defoliate forests in Vietnam—are used on lawns and other green spaces in the United States; the product can be purchased without a license in any hardware store.

Every day, the United States produces or imports 42 billion pounds of synthetic chemicals, 90 percent of which are created using oil. Converted to gallons, this volume is the equivalent of 623,000 gasoline tanker trucks, each carrying 8,000 gallons, which, if placed end to end, would reach from San Francisco to Washington, D.C., and back. In the course of a year, this line would circle the earth eighty-six times at the equator. Whether all that stuff ends up getting turned into consumer products, or burned, or buried, it never disappears. It has to end up somewhere. And it does.

It ends up in us.

Here's the problem: most of the tens of thousands of chemicals that are used commercially have been around for only a few decades, far too short a time for researchers to figure out with any certainty what impact they might have on our health. The human immune system has evolved over millennia to combat naturally occurring bacterial and viral agents. It has had only a few decades to adjust to most man-made contaminants, many of which are chemically similar to substances produced naturally by our own bodies. The effects of this are far from fully understood. "We face an ocean of biologically active synthetic organic compounds," the ecologist Sandra Steingraber writes. "Some interfere with our hormones, some attach to chromosomes, some cripple the immune system, some overstimulate certain enzymes. If we could metabolize them into benign compounds and excrete them, they would be less of a worry. Instead, many accumulate. So they are doubly bad: they are similar enough to react with us, but different enough not to go away easily."

To be sure, the scientific evidence linking toxic chemicals and health problems is still in its early phase. "There's not enough [information] to allow for big generalizations," Dr. Mark Miller, director of the Pediatric Environmental Health Specialty unit at the University of California, San Francisco, has said. "What it does do is show the huge need for this information, both to allow us to put these results in context and also to give us information on what's going on out there over time and over age groups. We're blind to what's going on out there."

In the spring of 2010, a major new study suggested links between toxic chemicals and autism—even if the exposure comes to babies still in utero. "If babies are exposed in the womb or shortly after birth to chemicals that interfere with brain development, the consequences last a lifetime," said the author of the study, Philip Landrigan, a professor of pediatrics at New York's Mount Sinai School of Medicine.

Today, despite vast sums spent on research and therapies, cancer continues to kill in increasingly devastating numbers. Lung, breast, and colorectal cancers remain among the leading causes of death among women; for men, prostate and liver cancers remain major concerns. One in three women will die of cancer. Among men, the number is one in two. Scientists estimate that roughly 34,000 people die each year from exposure to carcinogens in the workplace or the environment. "We are treating people like experimental animals in a vast and largely uncontrolled study," writes Richard Horton, a physician and the editor of the British medical journal the *Lancet*. "By any common-sense measure, the war on cancer has been lost."

And while cancer seems to strike the loudest bells in our collective consciousness, it is hardly the only danger that can be traced to environmental chemicals. Cognitive development experts say that learning disabilities rose 191 percent between 1977 and 1994. The California Department of Developmental Services says it saw a 210 percent jump in autism rates in the decade following the mid-1990s. One in eight children is born prematurely; nationwide, premature births have jumped nearly 30 percent since 1981. Twenty-four million Americans have an autoimmune disease; research indicates that this number has been doubling and tripling around the globe. The University of Kentucky recently reported a link between trichloroethylene, an industrial solvent known as TCE, and Parkinson's disease. TCE can be found in more than a third of the nation's waterways.

The deeper one looks into the spread of synthetic chemicals, the more fascinating, and frustrating, the story becomes. In 1984, the National Academy of Sciences completed a four-year study of high-volume industrial chemicals and found that 78 percent had not been subjected to even minimal toxicity testing. More than a decade later, the Environmental Defense Fund found "no significant improvement." And these were for chemicals produced or imported at volumes exceeding a million pounds a year, chemicals that were *already* considered environmental or health threats.

The more I began to look into this, the clearer it became that we have spent our lives virtually marinating in toxic chemicals: in the water that comes through our tap; in the plastics we find in our baby toys or use to store our food; in our soaps and shampoos and cosmetics; in the fabrics and dyes in our clothes and our furniture; in the products we use to clean our homes; in the chemicals we spray on our weeds and apply to turn our toilet paper white. Like everyone else, I suppose, I'd had moments in my life—when I used my bare hands to roll out insulation, or breathed in a neighbor's lawn chemicals, or tried to work in an office as the hallways were being painted—when I'd wondered whether the stuff could somehow be "bad" for me. But I never really bothered to think too hard about it. Honestly, I figured, if a product is sitting on a supermarket shelf in the United States at this late date in history, how bad could it be? There was scientific research behind all this stuff, and ethical business practices, and scrupulous government regulation. Wasn't there?

What becomes clear, if you stop to think about it, is that what's gotten into us is not just chemicals but culture. We aren't just saturated with chemicals, after all; we are saturated with products, and marketing, and advertising, and political lobbying. Fifty years ago, it was not uncommon to see advertisements for DDT featuring an aproned housewife in spike heels and a pith helmet aiming a spray gun at two giant cockroaches standing on her kitchen counter. The caption below reads, "Super Ammunition for the Continued Battle on the Home Front." Another ad shows a picture of a different aproned woman standing in a chorus line of dancing farm animals, who sing, "DDT is good for me!" DDT was marketed as the "atomic bomb of the insect world," but also as "benign" for human beings. And we believed it.

Our ignorance is not an accident. We are not meant to know what goes into the products we use every day. The manufacturers of most American-made products tend to keep the ingredients and formulations of their products secret, and rarely mention that individual ingredients might (or do) cause cancer, or impede fetal development, or lead to hormone imbalances. It seems that the intention in packaging is to make information harder to find, not easier—an imitation of information, not information itself.

Those "ingredients" lists on cleaning products or lawn chemicals—can anyone pronounce those words, let alone figure out why they're there? My dentist keeps a big box of disposable gloves in his examination room. The box says the gloves are made of polyvinyl chloride, which is known to cause leukemia and suspected of playing a role in bone marrow cancer and non-Hodgkin's lymphoma. Sure, the gloves protect him from patients' blood. But do I want PVC in my mouth? One day, when I looked inside my new Toyota Prius, I found a note saying that the car's air bags, batteries, and seat belt components may contain perchlorate, and that I could find more information on the compound by visiting the website for the California Department of Toxic Substances Control. When I did this, I learned that perchlorate interferes with the thyroid gland and is "becoming a serious threat to human health." So much for driving "green."

Trying to understand a list of ingredients is like trying to understand the super-sped-up disclaimer at the end of a radio commercial for a new drug. What was that the guy said about side effects? If such information is "presented" but is functionally unintelligible, what is the reason for including it? This gesture at providing packaging "information" seems ethically compromised, at best.

Pick up a can of organic tomato sauce, and you may read that the can is lined with "lead-free enamel." Sounds good, right? But what the can doesn't tell you is that the can is lined with bisphenol A, a plasticizer that has been shown to cause hormone imbalances; this information is available only if you put in a phone call to the company. Should the plastic be listed on the can? Should the presence of bisphenol A affect the "organic" label itself?

Similarly, "spring water" is bottled in a container that is made with phthalates, a plasticizer that may lead to lower sperm counts and that has been shown to leach from bottles into the liquids they contain. Should that water really be considered "pure"?

"Information," such as it is, is hard to come by, harder to understand, and harder still to weigh. Again, this is no accident. Efforts to force companies to list more—and more understandable—information on their products have run up against stark political reality. Chemical companies have contributed $47 million to federal election campaigns since 1996 and pay $30 million a year for lobbyists in Washington. The industry uses its clout at both federal and state levels to kill most efforts at increasing what we can know about these toxins. Is it any surprise, then, that most chemicals have never been even minimally scrutinized for their toxicity?

Granted, part of this is our own fault. When we buy a box of "chicken nuggets," part of our brain registers the distance this "food" has traveled since it—or a small part of it, anyway—walked the earth as a living creature. Beyond this, we convince ourselves not to ask too many questions. The same is true for most of the products discussed in this book. We don't want to know about toxic plasticizers, we just want a bottle of water. We don't want to know about threats to our fertility, we just want a tube of lipstick. We don't want to hear about breast cancer, we just want to buy a new couch. And so on.

With so little information, it's easy to see why we have become so complacent. And why we have allowed ourselves to live, albeit uncomfortably, with assurances that these products are "safe." A single exposure to these chemicals never killed anyone, we tell ourselves. This is true. But smoking a single cigarette never killed anyone, either. The trouble with exposure to toxic chemicals, as with exposure to tobacco, is that the impact is cumulative, long-lasting, and, frequently, slow to reveal itself. Flame retardants may be in your mattress, but they are also in your children's pajamas and in the salmon you had for dinner last night. If one exposure to bisphenol A, the plasticizer found in many plastic bottles, is considered "insignificant," how about a hundred exposures? A thousand? How about twenty-five years' worth? And what happens when you are exposed to bisphenol A in combination with hundreds of other chemicals over these same twenty-five years?

"The effects don't just accumulate, they mushroom," writes Devra Davis, the director of the Center for Environmental Oncology at the University of Pittsburgh Cancer Institute. "Scientists have long known that certain chemicals . . . can cause cancer. Now we're beginning to realize that the total of a person's exposure to all the little amounts of cancerous agents in the environment may be just as harmful as big doses of a few well-known carcinogens. There's plenty of evidence that combined pollutants can cause more hard harm together than they do alone."

More questions: What role will these chemicals have on our pregnancies, or our children? Which of them might affect us in old age? The truth is, we don't really know. Except for leukemias, most cancers don't show up until three or four decades after an exposure, and even then they sometimes show up only indirectly. Lung cancer and the myriad other diseases caused by smoking are caused by repeated exposure over a stretch of time. When it comes to how a body will respond to toxins, who can say? By the

1950s, death certificates were already showing a far greater proportion of people dying from cancer than at the turn of the century, and pediatric cancers were becoming commonplace. Women born in the generation just before mine—between 1947 and 1958—have almost three times the rate of breast cancer as their great-grandmothers. Since then, the presence of chemicals in consumer products has vastly increased. For starters, over 5 billion pounds of pesticides had been used worldwide by 2001—one-quarter of them in the United States.

More recently, experts say, breast cancer, on average, robs female victims of twenty years of their lives, which means one million years of women's lives are lost every year. This was true for Rachel Carson, who died of breast cancer in 1964, at the age fifty-six. She had been diagnosed in 1960, in the middle of her research for *Silent Spring*, and though tumors made her writing hand numb and radiation treatments left her nauseous and weak, she continued working. Eighteen months after *Silent Spring* was published, she died.

So here we are.

Almost fifty years after *Silent Spring*, and the tide of synthetic chemicals is only rising. We are faced, every day, with an overwhelming number of choices as consumers: Do I choose this detergent or that one? This mattress or that one? The chemical lawn-care company or the "green" one? This shouldn't be so hard. We're talking about washing our children's hair. Or cleaning the sink. Or tending a garden. Why has this become so complicated? And on what information do we make our decisions?

That is what this book will try to address. The most important thing, as I have said, is finding the courage to see things clearly. But as I have learned, when it comes to toxic chemicals, seeing things clearly is harder than you might imagine.

In every room of our houses, in every action we take, we are exposed to synthetic chemicals. You hear that virtually every stream in the United States has been found to contain hormones found in oral contraceptives, so you decide tap water isn't such a great idea. You buy spring water in plastic bottles, only to learn that the bottles leach chemicals into the water; even if the water is clean, the bottles aren't.

We inhale chemicals in the form of "volatile organic compounds" off-gassing from paints, perfumes, and synthetic upholstery. We put them on our skin, in the form of cosmetics, moisturizing creams, and shampoos. We absorb them in our food and in our drinking water, which is laced not only with agricultural pesticides but with discarded pharmaceutical drugs like antipsychotics and erection enhancers. Once you start to peek behind the curtain, the number of chemicals we expose ourselves to becomes unnerving. If we can even stand to peek, that is. Most of us are so numb or frightened by this saturation that we don't have the courage to ask where all this stuff comes from or what it might do to us—far less, what we might do to pull ourselves out of this stew.

So what actions can we take on our own to protect ourselves from—or, at the very least, limit our exposure to—these toxic chemicals? What can we do around our homes, and our workplaces, and our children's schools? What should we talk to our neighbors about, and to our legislators? How can we regain some sense of control over what ends up in our bodies?

Over the years, we have delegated so much knowledge about (and so much responsibility for) our health to "professionals" who grow our food, make (and inspect) our toys, "manage" our lawns, "treat" our water. When even the simplest things are outsourced to people whose interest in the process is professional and, therefore, economic, what surprise is it that shortcuts are taken? When someone at a chemical company says, "We are not aware of the dangers of these chemicals," who exactly is "we"?

Every choice we make is a bargain with the devil. You go to get your suits dry-cleaned, only to learn that dry cleaners rely on perchloroethylene, or perc, a known carcinogen. Is having crisp creases worth the risk? You want to wash your infant's hair. What could be more benign than baby shampoo? But look closer at the label on the bottle: the baby shampoo contains formaldehyde, which causes cancer and compromises the immune system. Formaldehyde in baby shampoo? Isn't that the stuff they used to preserve dead frogs in high school science classes? In baby shampoo?

How about Christmas trees? Which would you rather have in your living room: a live tree from a tree farm that has been sprayed with pesticides or an artificial tree made with polyvinyl chloride, which is made from petroleum and emits dioxins when burned in an incinerator?

The more alienated we get from the things we use every day, the more confused we get. The more confused we get, the dumber we feel. The dumber we feel, the less confident we are in our decisions. The less confident we are, the more susceptible we become to the suggestion that everything is as it should be, that the experts (the manufacturers, the regulators) are keeping an eye on things. The more we bury our worries under such shaky ground, the more abstracted we become. Ask yourself the simplest possible questions: Where does your drinking water come from, how safe is it to drink, and how do you know? Can you answer these questions with genuine confidence? Now ask yourself the same thing about something less simple, like your eyeliner, or the glue in your plywood, or the rubber duck floating in your child's bathtub. Where do these things come from, how safe are they, and how do we know? And finally, think about this: Even if you have confidence in American regulation, what do you say when you learn that $675 billion worth of consumer products—everything from pet food contaminated with melamine to children's train sets made with lead paint—are imported from China, where government oversight is, shall we say, rather lax?

As overwhelming as some of the scientific evidence about our consumer products can seem, there can be real liberation in learning to look at things with clear and unblinking eyes. A good part of this has to do with reconnecting with our things. With understanding what things are for, and how they are made, and by whom. I've come to think of this as "Learning to Live Like Your Grandparents Did," provided that your grandparents grew up before World War

11. Since the vast majority of toxic synthetic chemicals were developed during or after the war, it stands to reason that people once managed to live reasonably well without them. We're not talking about prehistoric man here. And without meaning to sound nostalgic, I think it's fair to say that, chemically speaking, those were simpler times. It's worth relearning some of what we've forgotten. How to build and furnish and clean our houses. How to care for our lawns. How to feed and clothe and bathe our children.

Strangely enough, you might find that some of these old ways actually feel empowering. We've been bombarded with advertising and marketing ploys for so many years that we have tended to make decisions out of unconscious habit rather than conscious choice. Not only is it increasingly clear that there is physical risk in such habits, there is also a genuine psychological sacrifice. In the moment when we reach unthinkingly for a product, we suspend judgment and even, at times, common sense. When we act unconsciously, we implicitly grant authority—and trust—to what manufacturers have told us, that a product is "safe." But the truth is, whether the product is an apple, a T-bone steak, a can of air freshener, or a mattress makes no difference: we have no idea what has gone into creating the product, even if someone, somewhere, has assured us that the product is benign. In many, many cases, this is clearly no longer true. And as the physical and psychological distance has grown between us and the products we consume, we have traded an intimate knowledge for a vague and anxious "trust," a feeling that is inevitably accompanied by its darker corollary, fear and loss of control. This does not seem like a fair trade.

In addition to chronicling our addiction to toxic chemicals, then, this book will also offer some thoughts on how we can begin to pull ourselves—and our children—out of this particular soup. As practical as some of the suggestions will be—what kinds of products to buy and which to avoid—they will also implicity urge that we develop new relationships with our material goods. Rather than mindlessly reaching for a product, we might . . . pause, just for a moment, and ask: What do I know about this product? *How* do I know what I know? Who told me? Can I believe what I have been told, what I think I know? A pause at that moment is critical, because it offers the chance to replace mindless action with conscious choice and genuine responsibility. No longer am I automatically, unthinkingly filling my shopping cart. Now I am mindfully choosing this, or choosing that, or choosing nothing at all. There's power in that pause. And reclaiming this power is energizing on both small scales and large—not least because it makes us realize how much we have given up, and how much we have to regain.

ONE The Body

To make a point about the saturating presence of toxic chemicals in the environment, field scientists will, on occasion, leave off looking for contaminants in big cities and abandoned industrial sites and travel to some of the world's most remote places. In recent years they have found petrochemicals—and breast cancer—in the bodies of beluga whales in Canada's St. Lawrence River. They have found PCBs—compounds used in electrical transformers that have been banned for thirty years—in the snow atop Aconcagua, the highest mountain in the Andes. They have even found flame retardants in the blubber of seals on Canada's Holman Island, far above the Arctic Circle. Synthetic chemicals, it turns out, circle the globe like the winds.

Despite such evidence—that toxic chemicals are, in essence, everywhere—human health advocates have struggled for decades to convince the public that there may be a link between so-called environmental toxins and individual and community health. After the stir caused by the publication of *Silent Spring* in the early 1960s, it took a full decade for the government to pass, and begin to enforce, pollution controls in factories and hazardous waste dumps. And thirty years after that, it remains more difficult than ever to convince people that the products they rely on every day—products that are made, after all, with these same toxic chemicals—might in any way be risky to use.

It's important to understand that your body is already full of toxic chemicals. This is true even if, as the saying goes, you were born yesterday. Long before you ever bought a flame retardant couch, or a sheet of plywood, or a can of ant spray, the chances are quite good that you absorbed toxins through your mother's placenta, her breast milk, or both. Given the ubiquity of chemicals in our lives, the accumulation grows from there.

In Maryland, where I live, a lot of attention is paid to the health of oysters, one of many endangered species suffering from toxic runoff in the Chesapeake Bay. Oysters spend their days on the bay floor, filtering water in one end and out the other. Whatever microscopic material is in the water passes through the oyster. Most of it exits; some of it stays inside. These toxins can be measured.

What is becoming clear is that we are all oysters. We are all exposed to all kinds of toxins. Some of these we filter out; others stay inside us.

In recent years, public health groups have come up with a new tactic to make this point: the body burden study. Such studies are not, at least primarily, invested in proving that toxic chemicals are "dangerous." This work is being done, with increasingly clear results, in scientific laboratories. What the body burden studies do is prove that these chemicals are everywhere—in the environment, in wild and domestic animals, and, with increasing frequency, in our bodies. Proving that toxic chemicals are dangerous hits people in their heads. Proving that people have chemicals *in their bodies* hits people in their guts. For decades, the chemical industry has been able to convince our heads that chemical harm is still in dispute, that "more research is needed." The authors of the new body burden studies are betting that the gut is less easily persuaded.

"Our experience with persistent chemicals of the past such as

DDT and PCBs has shown what happens when we wait to gather conclusive evidence of a chemical's harm instead of acting on mounting evidence," the Public Interest Research Group reported in 2003. "By the time the chemicals were regulated, they had spread across the globe and left a path of damage from which we have yet to recover."

If lab science aspires to prove chemical harm, body burden studies aspire to show chemical exposure. In Europe, linking harm in the lab with exposure in the community has been enough to prompt radical changes in the way toxic chemicals are regulated. "In a court of law, a person is innocent until proven guilty," a United Nations report on the persistence of environmental toxins says. "Chemicals suspected of bio-accumulating, persisting in the environment, and harming human beings and animals do not deserve that kind of protection. Unless precautionary action is taken to curtail exposure to these chemicals, millions of people—not to mention millions of other creatures ranging from lake trout to penguins—are likely to suffer terrible harm." As of three years ago, chemicals in Europe are considered guilty until proven innocent. Here in the United States, it is still the other way around.

When I wanted to find out how ubiquitous synthetic chemicals had become in people, I decided to go to Maine. I wanted to meet some folks whose bodies, I had heard, had recently been tested and found to be full of plasticizers. And mercury. And stain resisters. None of these people worked in a laboratory, and they had not grown up in big industrial cities, or near the chemical corridors of Louisiana, Houston, or Delaware. One was a woman raising young children in rural western Maine. Another was a twenty-eight-year-old woman raised in one of Maine's remotest corners. A third was an organic farmer.

How did this happen?

Lauralee Raymond grew up in Fort Kent, way up in Aroostook County, near the St. John River, on the Canadian border. This part of Maine is a paradise of rivers and lakes, where moose can seem to outnumber people and canoeists from all over New England ply one of the East Coast's great remaining wildernesses. Lauralee's family has lived in the north country for generations: her father's family is from Acadia, her mother's from Quebec. Lauralee's great-grandfather and grandfather were both potato farmers. The family, for a very long time, has been connected to the land.

Ask a Maine native what they consider to be northern Maine, and they are likely to say, "Bangor." But you'd need to continue another four hours north from Bangor to get to Fort Kent. "If you drive through Aroostook County, there's so much forest, you don't see houses forever," Lauralee says. Since Interstate 95 stops two hours south of town, the only way to get to Fort Kent is to follow a winding road cutting through the northern woods. There are moose and deer at every turn, Lauralee says, and if you want a true adventure, drive the road at night.

Surrounded by such a wealth of natural beauty, Lauralee spent most of her childhood outside. She and her friends swam in creeks. They rode their bikes and skied cross-country. Every fall, they got a few extra weeks off to help with the potato harvest—which, for the kids, meant separating the potatoes from the rocks and the mud. Nowadays the area is known mostly as an Olympic training center for biathletes—who combine cross-country skiing and shooting—and for hosting a qualifying sled-dog race for Alaska's Iditarod.

As she grew older, Lauralee moved downstate to attend Bates College, in central Maine, then settled first outside Augusta and later in Portland, where she now works for a women and children's policy group. When I met her, she was sitting in a coffee shop that was festooned with signs encouraging customers to support Maine's economy by buying local. Local food. Local music. Local beer. In one corner, a trio of women sat knitting. In another, a woman nursed a child in her lap. This, it turned out, was Lauralee's kind of place. She is a cheerful, open-faced, energetic young woman, and fiercely proud of her state's eccentricity, its rural character, and its independence.

But in recent years Lauralee has had this sensibility shaken. She had agreed to meet me in the café to talk about a study in which she had taken part that had made her question a great deal about her ability—and her state's—to exist apart from the corrupting influences of the urbanized world. A couple of years ago, she had participated in a study being conducted by a public health group hoping to draw attention to the growing presence of toxic chemicals in everyday consumer products. Each participant would donate samples of their hair, blood, and urine to a research team from the Harvard School of Public Health and the University of Southern Maine. Once the samples were collected, they would be sent to laboratories in Seattle and Victoria, British Columbia, for chemical analysis. Technicians would test the samples for a spectrum of toxins. The lab would not offer diagnoses; there would be no attempt to link contamination to current or prospective diseases. All the volunteers would learn was what they had lurking in their bodies.

Thirteen people agreed to participate. They came from all walks of life: A furniture store owner. A teacher. A nurse. They were men and women, young and old. Several represented a group that has become a very important constituency for public health advocates: they were women of childbearing age. At twenty-eight, Lauralee Raymond definitely qualified.

Public health advocates hoped that proving the presence of toxic chemicals in a randomly selected group of citizens would cement the notion that toxic chemicals were more than a problem limited to people who lived near Superfund sites, or showed up only in residents of New Jersey. The study's sponsors were interested in research and advocacy in equal measure: if people in Maine were contaminated, the thinking went, people everywhere were probably contaminated, and something ought to be done about it.

Before the test, Lauralee Raymond was confident, even cocky, about the relative purity of her body. She was in her twenties. She was a runner. She ate organic food. She had spent her childhood in one of the most pristine corners of one of the most rural states in the country. If anyone's body was clean, she figured, it would be hers. To make matters more interesting, her mother agreed to be tested as well. When it came to the number of chemicals in their bodies, Lauralee felt certain that youth would be served.

"I went into this as a kind of game, or a competition with my mom," Lauralee said. "I was like, 'Oh my gosh, this will be fascinating, to see how much better health I am in than my mom.' Look, my parents can't even pronounce 'tofu.' I figured, maybe this will get my mom to take better care of herself."

The organizers of the study had told her, and all the participants, that they should be prepared for a few surprises. Lauralee scoffed.

"I was thinking, like, 'You don't need to tell me that. I'm going to be fine. My results will be fine.'"

Russell Libby was also raised in a rural part of Maine. He grew up in Sorrento, outside Bar Harbor. His people were from modest means, mostly farmers or retailers; his grandmother was a drugstore clerk, his father a state trooper. As a kid, Libby worked at a golf course where they used a lot of pesticides, and then worked raking commercial blueberries, where they used still more. Later, he was trained as an economist at Bowdoin College and the University of Maine. Libby has devoted his career to studying agricultural policy. He spent a decade as research director of the state's Department of Agriculture, and has been involved with the Maine Organic Farmers and Gardeners Association for thirty years, the last fifteen as its executive director.

He is also a published poet. Like Wendell Berry, Libby is given to lines reflecting the pressures he, and his rural state, feel from the industrialized world. One poem, "Worth," opens with a quote from a vice president of Dow AgroSciences recommending that farmers work a piece of land "for all it's worth." The next stanzas raise questions that shift our attention from the economic to the metaphysical:

For how many bluebirds it's worth?
For how many monarchs?
What price the elusive fireflies?

I pulled the early peas today,
tossing the vines in the compost bin,
then carried the sack of Tartary Buckwheat from the barn,
seed grown by Liz and Chris on their farm,
and sowed it in the same way
farmers have sowed since the beginning,
palm up,
fingers pointing in the direction the seeds are thrown.

And what is that worth?
To hear the seeds meeting the ground,
to look up and see the clouds
that will bring rain tonight or tomorrow,
and know next week the ground will be covered
with pale green, triangle-shaped leaves,
six weeks before the white flowers will carry bees.

I met Libby at the Maine Agricultural Trades Show in the middle of January. The civic center had everything you'd want in an agricultural trade show: Raffles for a new tractor. Maple-sugaring equipment. Sign-up booths for the Sheep Breeders Association and the Beekeepers Association. There were portable sawmills and a man demonstrating the strength of a plastic shovel by jumping on it. This thing is so strong, he was telling a customer, that he once ran over it with his truck and it just snapped right back. Off in one corner, I picked up a pamphlet that promised to teach me how to build an outdoor oven out of clay.

It was here that I figured out what Maine farmers grow in winter: beards. I have never seen such a fine crop in my life. And I'm not talking about the neat goatee you might find on a Boston banker, or the sideburns-and-soul-patch combination you might find on a bartender in New York. This is Maine. I'm talking about full bushes, capable of warming not just chins but chests.

I met Libby at a booth set up by the Maine Organic Farmers and Gardeners Association. His lively blue eyes lit up—naturally—a thickly bearded face. He wore a green corduroy shirt and hiking boots, and, judging from the number of people who came by to

shake his hand, he seemed to know everyone. We walked over to a set of folding chairs beneath a podium used for product demonstrations and sat down to talk.

Like Lauralee Raymond, Libby told me he had spent the bulk of his life in rural communities, and had been unusually conscious in his choices. He hasn't used synthetic fertilizers or pesticides on his farm since he started it twenty-five years ago. When it came time to build a house, he and a neighbor "made a whole lot of decisions to take things out of play." They used only native woods, doubling up layers rather than using plywood, which he knew contained formaldehyde. They didn't lay carpeting, since most of it contained toxic flame retardants. The whole project, he said, was "very low-tech."

If Lauralee Raymond figured she'd be clean because of her youth, Russell Libby figured he'd be clean because of his choices.

The Maine study was modest in scale. It set out to test for a handful of chemicals known to be toxic: phthalates, softening plasticizers that are added to everything from baby toys and plastic shower curtains to drinking-water bottles and soft lunch boxes; flame retardants, which are mandated in everything from television sets to sofa cushions and draperies; perfluorinated chemicals, the so-called Teflon treatments that are used as stain-resistant coating on furniture upholstery and as nonstick coating for cookware and food packaging for things like microwave popcorn; and bisphenol A, a hardening plastics additive used to make baby bottles and the lining of food cans. The study would also look for a handful of dangerous metals like lead, mercury, and arsenic, which are found in everything from old paint to new Chinese-made toys, in power plant emissions (and the fish contaminated by them), and in pesticides once used in pressure-treated lumber and on roadways and golf courses.

The chemicals and metals were chosen from tens of thousands of others in wide use today in part because they are so common, in part because some of them are so long-lasting, and in part because a growing body of science is warning that they may be dangerous even at very low levels. The researchers were quite clear about the mysteries involved. Finding evidence of individual chemicals in the bodies of thirteen people would not necessarily allow them to predict that any of the study's participants would ever develop an illness. Such direct correlations, at least at this stage, are virtually impossible to establish. While some of these chemicals (lead, arsenic, mercury) have been around forever, others (phthalates, bisphenol A, flame retardants) are relatively new formulations. Scientists, after all, have been looking into some of these chemicals for only a few short years, and are just beginning to understand their effects on human health. How do chemicals migrate from consumer products into the environment? Once in the environment, which pathway do they follow into the body—through the nose, mouth, or skin? Once in the body, how do they combine with other synthetic chemicals? What impact do they have—alone and in combinations—on our health and on the health of our developing fetuses and babies?

The truth is, we are only beginning to make connections between exposure and illness. "There is much that we do not, but should, know about a large number of widely used chemicals in household products," writes Richard Horton, the editor of the British medical journal the *Lancet*. "There seem to be few incentives to study these chemicals and risk to human life. A 'medical surveillance' program to monitor these potential hazards is a good idea. There are too many vested interests in medical research, deflecting scientists from studying important questions of public health. And we do need to discover new ways to defeat cancer."

Given the immense chemical and biological complexity of the human body, tracing a single chemical exposure to a precise health problem is next to impossible. A host of factors influence whether or not one's exposure to toxics will lead to illness, including the nature and concentration of the chemicals, at what point in life a person was exposed, and how often and for how long. A person's own genetic makeup is also important, as is their general health and their socioeconomic status. A lot of variables, in other words, both in the chemicals and in the bodies they contaminate. Nonetheless, health organizations, both local and federal, are beginning to agree. "It is the concentration of the chemical[s] in people that provides the best exposure information to evaluate the potential for adverse health effects," the federal Centers for Disease Control reported in 2005.

The Maine study was not the first of its kind. In one 2007 multistate sampling of 35 people from Massachusetts to Illinois to Alaska, every participant tested was found to be contaminated with flame retardants, and all 33 people who donated urine were found to harbor phthalates and bisphenol A. In 2007, a "Pollution in People" study of 10 Oregon residents tested for 29 toxins and found 19, with an average of 12 per person. In 2005, an examination of "the pollution in newborns" found some 287 industrial chemicals—including 180 that cause cancer, 217 that are toxic to the brain, and 208 that can cause birth defects or abnormal development—in umbilical cord blood taken from ten babies around the United States. The blood samples, initially collected by the Red Cross, contained "pesticides, consumer product ingredients, and wastes from burning coal, gasoline, and garbage."

The CDC began its own large-scale body burden study in 2001, and has updated it every few years since. The first year, researchers tested people for 27 chemical compounds. Two years later, the agency tested for 116 compounds. In 2005, the number of chemicals tested for rose to 148. The goal of the study, the CDC reported, is to "determine which chemicals get into Americans and at what concentrations." For chemicals with known toxicity levels, researchers hoped to determine "the proportion of the population with levels above those associated with adverse health effects" and to see if there were groups with unusually high exposures, especially minorities, children, and women of childbearing age.

The CDC's most recent update noted a few encouraging signs, which are apparently the result of increased government regulation: exposure to toxic secondhand smoke appears to be dropping, as does childhood contact with lead, though children living in older buildings continue to be "a major public health concern" and

African-American children have twice the chemical load of white children. Several pesticides, some banned for twenty or even thirty years, are also beginning to show up at much lower levels.

But there is still plenty to worry about.

Exposure to phthalates, the CDC reported, is "widespread." Synthetic pyrethroids, a common type of insecticide, were "found in much of the U.S. population." Yet the overriding sense from the CDC report was a kind of scientific (and therefore political) ambivalence. For most of the environmental chemicals in this study, the CDC reported, "more research is needed to determine whether exposure at levels reported here is a cause for health concern."

D. Richard Jackson, the former director of the CDC's National Center for Environmental Health, which oversees the national biomonitoring project, came under fire from industry critics. "I took a fair amount of criticism for disseminating the report without putting it through some kind of extensive risk assessment," he said in March 2004. "But I resisted that very strongly, not because I was antiscientific, but because I wanted the larger community and the research community to have it in their hands and use this data the way a doctor would use lab data in making decisions about a patient. The complaint from chemical manufacturers was that the report was just going to scare people. No, you never scare people with real information. You scare them with no information or bad information."

When it comes to scientific debates, few phrases are more fraught than "more research is needed." This expression, in the mouth of an independent scientist, can mean two things: "I have come up empty—more research is needed" or "I have just made a significant discovery—more research is needed." In the mouth of an industry spokesman or an industry lawyer the expression can mean something entirely different: "We don't believe existing science proves that our product causes harm—more research is needed." In other words, depending on its source, "more research is needed" can reflect diametrically opposed impulses. One is an impulse to raise a red flag; the other is an impulse to take a red flag down.

This all sounds fine when you're talking about abstract science. But when it's your phone that rings, and it is a physician on the other end of the line, and he is telling you that your body is full of toxic chemicals, such abstractions can seem irrelevant. Just ask Lauralee Raymond. When the doctor running the Maine body burden study called her up to discuss her results, he presented a picture of her cells that was rather at odds with the image she had of herself as an uncontaminated girl from the north country.

Lauralee was not the only Maine volunteer who was shocked by her test results. Researchers had examined thirteen people, none of whom lived in major cities, none of whom worked in heavy industry. They'd gone looking for 71 toxic chemicals. They found 46. On average, each person in the study harbored 36 different toxic chemicals in their bodies. Every person had detectable levels of lead, mercury, and arsenic. All had traces of the chemicals used to make Teflon cookware and Scotchgard treatments for stain-resistant fabrics—even though the latter have not been produced since 2001. Three women had blood levels of bisphenol A that were six to ten times higher than the national average. The highest level—ten times the national average—was found in an eighteen-year-old girl named Elise Roux. All five women of childbearing age had levels of mercury that were higher than the national average. The levels of the phthalate used in hairsprays, insect repellants, and soft plastics were higher than in 95 percent of Americans tested in a national study. Three other phthalates, used in PVC products like auto interiors and shower curtains or in vinyl flooring, nail polish, and other cosmetics, were higher in the study participants than in 75 percent of all Americans.

Hannah Pingree, a young state legislator from Down East, had recently gotten married and was planning to have children. When she got her results, she learned that she had mercury levels above the standard for protecting a developing fetus from "subtle but permanent brain damage." Dana Dow, a furniture store owner and former Republican state senator from Waldoboro, had the highest levels of perfluorinated chemicals (PFCs) and Teflon compounds in the test group. For a number of toxins, Senator Dow's levels were more than twice the national average.

"Maine people are polluted with dozens of hazardous chemicals," the study reported. "These chemicals are found in products we use every day: plastic containers, toys, furniture, fabric, TV's and stereos, water bottles, medical supplies, and personal products like shampoo, hairspray, and perfume. They are in our homes and offices, food and water, and the air we breathe. That 'new car smell'? It's a collection of volatile organic compounds off-gassing from flame retardants, glues, and sealants.

"What is most unsettling for the project participants," the study's authors went on, "is that no one, not even the doctors leading the study, can explain with any certainty why particular chemicals were found in their bodies, why levels of some chemicals are higher than others, or how the chemicals are affecting their health now or in the future."

So how *were* all these people exposed?

Dana Dow told researchers he often used furniture sprays in his shop and often brought treated furniture home. Maybe that was it. Bettie Kettell, a surgical nurse, who had the highest total number of flame retardants in her body, worked in a community hospital that, like any commercial building, has higher fire-safety standards—and thus, presumably, more flame retardants—than private homes. Her hospital had recently been outfitted with new rugs, drapes, and furniture. Like any nurse working in a hospital, she is also exposed to countless computers and monitors, most of which are made with flame retardants in their plastic casings.

But it wasn't just Bettie Kettell who was loaded with flame retardants. All thirteen participants were. Including Lauralee Raymond and Russell Libby.

Lauralee Raymond had higher levels of many toxins than her fifty-one-year-old mother, despite twenty-three fewer years of exposure. She had high levels of mercury and arsenic. Somehow, as bad as these chemicals sounded, Lauralee was able to put them in a box in her head. Mercury, she figured, she got from eating so much fish. Okay, she thought. She'd give up sushi. No big deal. Arsenic? She had heard that arsenic was a common and naturally occurring mineral in the soils of Maine. Not much she could do about that. Maine was where she lived.

But flame retardants?

Lauralee said her results were especially rattling given what she had thought was a life spent making consciously healthy consumer choices. "We couldn't figure out where this stuff came from," she said. "It was really disturbing. We thought maybe it may have been a futon I bought a few years ago. It was the only thing I had purchased, that I had bought new, that may have had flame retardants in it. I had hardwood floors everywhere, I didn't have any rugs. I really hadn't bought very much."

The study's lead physician made it clear to Lauralee that given her age—and her desire to have children—she was at a particularly vulnerable stage. Not just in terms of new chemical exposures, but because of the load of chemicals she's accumulated over her young life. "He kept saying, 'Childbearing age, childbearing age,'" Lauralee said. "That's what's so astounding, how dangerous these chemicals are how and long they stay in your body.

"You kind of go through these phases. You're in denial, then you get freaked out. Then you go through a hypervigilant phase. I got rid of my microwave, got rid of my nonstick pans. I was really hypersensitive about what was going into my body. Then I got really pissed off. My mom and I went to a number of panels; people kept asking us what it was like. I always ended up getting really angry. I don't think it should take us five hundred years to get the science together. It's only been sixty years that these chemicals have been around? Sixty years seems like a very long time to me. That's what's so hard about this stuff. People always say, 'Show me the science. Show me the damage. Show me the evidence.' People always put it back on you, on the individual. People come up to me and say, 'Oh my God! I read about you! You're toxic!'

"'Well,' I always think, 'if I'm toxic, you're toxic, too.'"

Russell Libby was also loaded with flame retardants. In fact, among all the participants, he was tied for the highest number of chemicals: 41 of the 71 tested for. He had the highest total number of flame retardants, and the highest level of individual flame retardants. Twenty years Lauralee's senior, Libby sounded resigned, even melancholic, about his results. The way he talked, he seemed less worried about his own health than he did about the cumulative effect of synthetic poisons on his community. He is, after all, an organic farmer who has spent many years trying to convince people that chemicals sprayed on farmland ends up in food, and then in our bodies. But in recent years especially, Libby's feelings on the issue have become distinctly personal. In the past twelve months, nineteen of Libby's friends had died, ten of them from cancer: Pancreatic cancer. Esophageal cancer. Prostate cancer. Three more have recently been diagnosed with cancer. Some deaths are to be expected when you're into your sixties, he acknowledges. But *nineteen?* That's a lot.

"And those are just the people who have died, just among the people I know," he said. "When I was growing up in the 1960s, it just wasn't very common to hear about cancer. Maybe that's because a lot of cases were missed, but to me the incidence seems to have increased greatly. It's to be expected that you may die of prostate cancer at eighty. But at fifty-five or sixty? That, to me, is the discussion we aren't having."

Maine is a pretty safe place, Libby said, but when it comes right down to it, there really *are* no safe places. The mercury produced by midwestern power plants can be found in every higher-level fish in Maine. At this point in our collective chemical history, he said, it's safe to assume that all of us—humans or fish—have been broadly contaminated. Proving this with high-tech body burden studies only confirms what many of us already suspect—and in any case is only another step in the march to reduce this contamination.

"It's so expensive to do what was done with this testing," Libby said. "It costs several thousand dollars per person. And all of us are walking around with another whole set of residues that this study didn't even attempt to measure. It's not my lifestyle that caused my exposure, it's the world's. I've spent my life growing good food for my family and friends, and that's the starting point. But we all have to say 'enough.'"

Libby had recently read Linda Lear's excellent biography of Rachel Carson, and he found himself pondering just how far we have, or have not, come since Carson died in 1964, just two years after the publication of *Silent Spring*.

"Here she is, dying of cancer, and she gets up to testify before Senator Kefauver's committee. She doesn't say anything about her cancer. What she says is, 'Don't we have a right to be safe in our homes?' Forty-five years later we still don't have a broad-based consensus that we have a right to be safe in our homes. It's all about 'acceptable risks' and 'trade-offs.' Why should we even be talking in those terms?"

As we spoke, a woman navigated the crowd at the agricultural fair to say hello to Libby. He turned away from me to give her a word and a hug. She walked away.

"It's our duty to everyone around to not live with those kinds of problems, both as individuals and as a society," Libby said, turning back to me. He nodded in the direction of the woman who had just left. "Her husband died of cancer two weeks ago," he said.

It's reasonable, of course, to question the value of a body burden study like the one done in Maine. What does one such small survey mean to anyone other than the few people involved?

The chemical industry chimed in forcefully to discredit the study's results. "We're living longer and better than we ever have, and I think that these sorts of studies make people worry when I don't see a reason for that level of concern," said Dr. John Bailey, a spokesman for the Cosmetic, Toiletry and Fragrance Association.

Even as they dismissed the dangers of their products, some industry spokespeople acknowledged the threat body burden studies can have on their marketing strategies. "It's the perception of the user that matters," said Bill Lafield, a spokesman for the Consumer Specialty Products Association, whose clients make an-timicrobial cleansers, floor waxes, and air fresheners. "Even though there may not be a real risk there is a perceived risk, and if we can increase the comfort level of the perceived risk, we'll do that."

Such rhetoric does not sit well with Lauralee Raymond. Like a number of the other test volunteers, Lauralee has spent a fair bit of time speaking about the growing threat of toxic chemicals in front of community groups. She says she gets weary of being challenged about the evidence that, to her way of thinking, exists right inside her body.

"There's always some in the crowd who are skeptical," she said. "It's like they are working for the corporations. 'Don't get in the way of progress!' they say. One gentlemen I'm thinking of kept saying, 'The study wasn't scientific! How can we say things need to change when this isn't evidence? You have all these personal stories. So what? What can it tell us? Not a hell of a lot.' He left the panel halfway through."

She added, "When you read about these chemicals, that's where the numbness factor comes in. Sometimes I just need to check out a bit. It gets really overwhelming to think about. I get tired of being the advocate. I get tired of being the knowledgeable person, just because I have been in this study. What people need to understand is that we *all* have this stuff in our bodies."

Before leaving Maine, I drove north to Farmington, a rural college town in the center of the state, to meet one more participant in the body burden study, a young mother named Amy Graham. On my way up, driving along the Androscoggin River, I marveled, as I always do, at the clarity of the air in northern New England. That was until I started smelling the sweet, slightly sulfurous scent of the paper mills, which drifted south and east through the towns of Jay and Livermore Falls.

I pulled my car over the snow in the Grahams' driveway and knocked on the door. Amy invited me in and offered me some tea.

On the kitchen counter were both books and bananas; near the sink lay a bookmarked biography of Richard Wright, research for Amy's own series of biographies about African-American authors, designed for middle school readers. Her husband, Bill, had long since left for work. A forester, he was off snowshoeing through the woods of central Maine, tagging trees for a sustainable lumber harvest. Outside, it was fifteen degrees.

The tea poured, Amy and I sat around her kitchen table and began talking about her life. A fit brunette in her late thirties, Amy had grown up in Weld, a town in western Maine with a population of a couple hundred people. Her father was a real estate agent, her mother a schoolteacher, but they also ran a small family farm. Amy remembered having a few chemicals around, but the produce they grew was mostly organic. Like anyone who grew up in rural Maine, Amy could always smell the paper mills. In her part of the state, the mills that line the banks of the Androscoggin are the area's biggest employers; what you smell up there, people say, is "the smell of money."

Amy went to college in western Massachusetts. When she met Bill, he was working in the music industry in Boston, but after the wedding, he quit his job and the two set off to hike the 2,200-mile length of the Appalachian Trail. It was a yearning for a connected life—physical, rural, close to the land—that led the couple back to western Maine to raise their family.

As had been the case with Lauralee Raymond and Russell Libby, Amy Graham's distance from big-city living did nothing to protect her from industrial chemicals. She had, for example, among the highest levels of flame retardants, a particularly troubling finding given the links between such fat-soluble compounds and developmental problems in infants—and the fact that they can be passed directly from mothers to their nursing babies. Amy had breast-fed her daughters, Phoebe and Sylvia, in part because of the well-known nutritional and health benefits of breast milk but also because nursing allowed her to avoid introducing commercial products into such moments of mother-daughter intimacy. When Phoebe was an infant, Amy's brother showed her an article about flame retardants in breast milk. The report started Amy thinking.

"This was concerning to me, because at the time I was beginning to worry about Phoebe's health," Amy said. "She had severe food allergies, and awful eczema. Her skin was cracked and bleeding, she was refusing to eat. I had never seen skin so bad. When my brother showed me that study about breast milk, I thought, 'God, even my breast milk, the one thing I'm able to give that's healthy and is not a foreign food, maybe that's making her break out. Maybe this is contaminating her. Who knows?' That got me galvanized—even before I got tested."

Around this time, Phoebe also developed asthma, and several of her playmates were diagnosed with autism. Amy was no more equipped than anyone else to connect individual chemicals to specific illnesses, but what seemed to be a growing number of diagnoses did give her pause. And the more she looked, the more she began to see toxic chemicals in every corner of her family's life: in the kitchen, home furnishings, cosmetics, everywhere. Aggravating her anxiety was her sense of powerlessness to even know what was in this stuff—let alone what harm it might cause. She remembers that one afternoon, when a well-meaning college student painted her toddler's fingernails at a children's festival, she "freaked out." She had learned enough about the myriad unregulated chemicals in cosmetics, and she didn't want them anywhere near her daughter's skin. The college girl gave her a quizzical look. Not her finest moment, Amy confessed, but what was she supposed to do?

Her body burden results "awakened an underlying fear," she said. "I now had a heightened sense that my house was full of potential hazards: the carpets, food, cleaning products—everything in my day-to-day environment now seemed suspect.

"When Phoebe started having all these health issues, I felt pretty vulnerable and angry that these toxins are out of my control," Amy said. "When you're a parent, when you think about the future of the world and the future of your child, you wonder: Are things going to get worse or better? It just blew my mind that I could grow up here in the 1980s and not know any kids with autism, and now all these friends of mine had kids that were autistic. It's hard to say what causes it, but this is really something that we need to look at. It makes you think about some of the decisions your parents and grandparents made, and question them. Why do I have toxins going through my breast milk? Someone made some bad decisions along the way."

After she got her test results, Amy started shedding. She got rid of her vinyl shower curtain—which had been made with PVC and softened with phthalates—and replaced it with one made of fabric. ("I'd kind of *liked* the smell of new vinyl," she admitted, "but I shudder to remember . . .") She looked more closely at the label on her "natural" shampoo and discovered that its "fragrance" was suspended in phthalates. Who knew there was plastic in shampoo? She tossed it. The beloved Nalgene water bottles she and Bill had taken on their Appalachian Trail trip? They had sentimental value, but she now learned they had been made with bisphenol A, which leached into whatever liquid the bottles contained. Out they went. She looked again at the little white holiday lights she always hung in the wintertime, to light her home during Maine's darkest months. The label said they contained lead. Out.

The more she learned, the more she purged. She started buying nontoxic detergents. Sitting at her kitchen table, I noticed a "green" household cleaner, made by the Vermont company Seventh Generation, on her kitchen counter. Is that what she meant by making better choices? I asked. Sure, Amy said. But she could do this product one better. She reached into a cupboard and showed me a small sack of laundry detergent that had been manufactured just down the road by a mom-and-pop company called O-Nature-L. The label said the stuff was made of sodium bromate, soda ash, coconut oil, clay powder, vinegar, and essential oils. It was biodegradable, safe for septic systems, and harbored no phosphates or sodium lauryl sulfate. Ah, I thought. Now, this is a local economy. Where else but Maine can you buy locally made laundry detergent? And why would this be so?

As word got out that Amy had become something of an expert on toxic chemicals, friends started coming around to ask her opinions about things in their homes. Is this safe to use? they'd ask. Will this cause my babies any harm? One friend asked if Amy had heard about a local woman who had accidentally killed a dozen exotic birds—parrots, cockatiels, lovebirds—simply by cooking pork chops in a nonstick pan. At low heat. Amy checked into the story. Sure enough, an examination done at the University of Maine revealed that fumes released from the pan had likely caused the deaths of the birds. A local newspaper noted an item on the website for DuPont, which makes the nonstick chemicals: "Because birds have extremely sensitive respiratory systems, bird owners must take precautions to protect them. Cooking fumes, smoke and odors that have little or no effect on people can seriously sicken and even kill birds, often quite quickly."

And these are the pans we *eat* from?

Since it was invented in the 1930s by scientists at DuPont, Teflon has been part of our domestic vernacular. It's not only on nonstick pans; it (and similar compounds) can also be found in grease-resistant pizza boxes, water-resistant clothing, and any personal care product listing ingredients that include the syllables "fluoro" or "perfluoro." An FDA study has found these substances in the greasy paper that lines microwave popcorn packages at hundreds of times the concentrations in nonstick cookware, and the intense heat of the microwave oven may cause the chemicals to leach into the popcorn oil. This turns out to be something of a concern, given that Americans consume 156 million bags of microwave popcorn a year.

Which makes these chemicals a hot topic of conversation among the mothers of central Maine.

"A lot of moms I talk to want to know more," Amy said. "Pretty much everyone can understand a mom's concern, that your house—the space you have the most control over—is actually one of the most toxic places. That is a really scary thought. You look at cancer rates and autism rates that our kids are facing—we don't know why they are skyrocketing. We owe it to our kids to look into that. I don't want my kids saying to me, 'Why do we have this legacy? Why were our parents sitting around enjoying their consumer lifestyle and not questioning it?'"

Just before I left, Amy showed me a handful of plastic dolls her kids recently received as gifts. A little farmer with aqua blue overalls. A little blond swimmer doll with a one-piece bathing suit. They were cute, Amy confessed. But what was in them? Since they'd been imported from China, the chances were reasonable that they had been decorated with lead paint. Since they were soft plastic, they'd almost certainly been made with phthalates, which, as they break down, can stick to household dust and be inhaled or digested, where they can compromise the functions of a body's hormone system. The trouble with phthalates is that they are designed, in part, to make plastic pliable—and thus ideal for a curious toddler's oral fixations. Ever since her test, Amy said, she had become much more conscious of all the plastic toys her young kids were holding and putting in their mouths. She had already either hidden the worst offenders in the back of closets or tossed them out completely.

But these little dolls had sentimental value. "I just can't bring myself to throw them out," she said. "They got them from their grandmother."

For Amy, the dolls have become talismans, reminders of forces in the world that are both present and invisible. "Just the fact that there are products on the shelves that aren't labeled, that we're bringing into our homes—we allow chemical companies to pour out these products without any labels at all. The whole 'Isn't chemistry great' thing that went on in the 1950s—what a mistake that was. Sure it did great things, but now we're paying the price with our kids and their health issues. We need to go back and rethink those decisions."

TWO The Home

Amy Graham's decision to inspect every inch of her home—room by room, armed with a large garbage bag—got me thinking. Trying to maintain a healthy body and raise healthy children in a house stuffed with toxic chemicals made no sense. That much was clear. Katherine and I had long ago committed ourselves to serving our family organic food, so that wasn't the issue. We never used pesticides on our lawn, so we had done what we could there, too (more on this later). But the more I learned about the ubiquity of toxins in consumer goods, the more I realized that when it came to a serious cataloging of the potential hazards in our home, I didn't even know where to start.

I decided to do what any home owner does when confronted by their own naïveté and general incompetence: I hired a professional.

This would not be like other home inspections, where you're trying to find out about the strength of the roof, or the age of the chimney, or the efficiency of the furnace. It wasn't the fortitude of the house I was interested in. I wanted to know what kinds of things might make my family sick. What things were safe, and what we ought to throw away.

So where should I turn? The yellow pages are full of people you can hire to test your home for lead paint, asbestos, or radon gas. As important as these problems are to address, they mainly involve the remediation of things you yourself have little control over: lead paint has been illegal for years; asbestos insulation was likely installed long before you bought your house; and radon seeps naturally through the soil. Your responsibility is limited to getting rid of these things. I wanted to talk to someone who could tell me how to make better choices of my own.

I was looking for a kind of guru, if you will. Someone with genuine intelligence and real training in seeing what to my eyes was invisible. Someone uninterested (and even distrustful) of the marketing claims of "cleanliness" with which we are all so familiar. I didn't want someone to tell me what products I needed to sterilize my kitchen floor, or what I needed to spray in order to get rid of the ants that marched in endless formation by my back door, or what I needed to "freshen" the air in my bathroom. It wasn't germs I was worried about. Germs I can deal with.

The way I figured it, having a toxicologist walk through your home would be kind of like stripping down in front of your doctor or spending time on a therapist's couch: whatever stories you've been telling yourself—how fit (or fat) you are, how well (or poorly) you eat, how much you don't (or do) drink—they would all sound a bit hollow when held up to the cool eye of clinical scrutiny. But as with a doctor or a therapist, I was also willingly taking the risk that the benefits of what I would discover would be worth the pain of what I learned. I expected to be surprised.

I decided to call a man named Albert Donnay. Donnay is a Johns Hopkins–trained toxicologist and environmental health engineer with a full salt-and-pepper beard, frameless glasses, and tight curls in his hair. A bright, irreverent man, Donnay has the wry demeanor of someone used to revealing things to people that they have always lived with but never really considered. He inspects homes the way a physician inspects bodies: he looks at something you know intimately and gently reveals the consequences of your behavior. For better and for worse.

Donnay has chemistry in his blood. His parents were both highly regarded research crystallographers, scientists who studied the atomic structures of solid crystals. Donnay's mother got a PhD from MIT; his father, from Stanford. They taught together for years at Johns Hopkins and McGill, where they shared an office and always insisted on having their desks facing each other. "Imagine," Donnay said. "You're working at your desk, you lift your head up, and there's your spouse. Every day. For decades!"

Albert grew up with people like Linus Pauling sitting around the dinner table, and the talk wasn't just about the miracles of science. It was also about the dangers. For Donnay, Pauling—who had already won one Nobel Prize for his work in chemistry and another for his work against nuclear testing—became a model for the ideal career: science in the service of public health. As he grew older, Donnay made a conscious effort to follow Pauling's path: in school and as a young professional, he split his energies between chemistry and anti-nuclear work. When he started his own nonprofit group, Nuclear Free America, he convinced Pauling to serve on the board.

But it was another, lesser-known concern of Pauling's that became Donnay's abiding interest: the detection of toxic chemicals in people's bodies and what impact these chemicals may have on their health.

"It was only later that I realized what a pioneer Pauling was in recognizing biomarkers of exposure, and what a successful advocate he became to public health based on these biomarkers," Donnay said. "His quintessential case was his campaign to have mothers send him their kids' baby teeth, so he could check them for strontium 90, a radioactive element associated with the fallout from nuclear weapons testing. He didn't ask moms for their teeth, he asked them for their *kids'* teeth. The kids got exposed from breast-feeding, and it ended up in their bones.

"The brilliance of Pauling," Donnay noted, "was that he could track patterns of exposure, engage a bunch of women, and then turn it all into a successful campaign to stop atmospheric nuclear testing. And that was a big part of the issue that finally got Kennedy to go after a nuclear test ban. When ranchers got up in arms over their poisoned cattle, that didn't do it. When soldiers got exposed, that didn't do it. But when it was children's baby teeth, that really pushed the issue over the top."

Donnay's interest in the darker side of chemistry, it turns out, is more than intellectual and more than political. It is also personal. He blames his mother's early death, from stomach cancer, on all the years she spent standing behind leaky X-ray tubes in her crystallography lab. So when Albert Donnay shows up at your doorstep to look for toxic chemicals, he wants you to listen.

When you invite a toxicologist into your house, there are a few things you ought to know. You think you'll curry favor by telling him how frequently you shampoo your wall-to-wall carpeting?

That's very nice, he'll say. Get rid of it.

No, not the shampoo. The carpeting. All of it.

That couch—the one you paid extra for because it doesn't stain? Toss it.

That air freshener that makes the bathrooms smell like a spring glade? Throw it away.

Ditto for the scented candles.

And the synthetic lavender laundry detergent.

And the dryer sheets.

And the ant sprays.

And the dish soap.

Get rid of it all.

For Albert Donnay, there's "clean," and then there's clean.

When he inspects your home, he walks around with his nose in the air. The first thing he will ask you to do is gather all your cleaning products, laundry detergents, bug sprays, air fresheners, dryer sheets—anything, in effect, that you can *smell*—and dump it all in a big garbage bag. Then he has you tie the bag shut for a month. Once the month is up, you open the bag and . . . inhale. You should be prepared to be shocked. Donnay calls the garbage bag "the most effective thing I have to convince you to improve air quality in your home."

Donnay started our inspection by sitting down at our kitchen table, opening up a folder, and, over a cup of tea, asking Katherine and me a series of questions. When was the house built? When had we moved in? When was the last major remodeling?

I was intrigued. Was the fact that our house was nearly seventy years old a good thing, or were old houses trouble? We'd lived here for seven years and had done a major renovation in 2005. That was good, right?

Hmmm, Donnay said, writing on his sheet. Tell me about this renovation.

Uh-oh, I thought. It had been ten minutes, and already I was flashing back to the public health researchers in the cancer hospital, and their questions about my toxicological history.

We added a new kitchen and bedroom to the house, I told him. We put down new flooring, painted the walls, the whole bit. New cabinetry. This was good, wasn't it?

Not necessarily, Donnay said flatly. He lifted his eyes and, for the first of many times during his visit, explained what he clearly felt ought to be common knowledge. (The fact that it isn't, of course, is what keeps his services in such demand.) Remodeling projects, he said, present one of the most acute and complex sources of exposure to synthetic chemical compounds. Plywood sheathing and cabinetry are glued together with formaldehyde. Fiberglass insulation, paints, stains, and caulk—they all contain harmful chemical compounds. Astonishing amounts of dust are released, much of it—airborne, on the floor, in the countless corners and crevices—containing thick loads of toxins. All of this stuff sticks around in your house for a long time. It circulates in the air. It settles along your baseboards. It clings to your clothes. You can find it in your dryer lint. It gets on your skin, and on your dinner plates. It rolls in and out of your lungs.

Anyway, tell me more about your renovation project, he said.

Well, I told Donnay, since our daughter had been just eighteen months old at the time, we had decided to move into an apartment while the work was being done.

Excellent decision, Donnay said, adding a note to his sheet. You don't want a toddler playing with cans of floor sealants or taking in lungfuls of all that debris.

He continued with his questions.

How often do you use indoor bug sprays? (That very week, Donnay pointed out, a study released by a cancer center at Georgetown University had found that patients with childhood leukemia had elevated levels of household pesticides in their urine.) How about outdoor pesticides? Herbicides? Do you use a professional lawn-care company, and if so, do you know what they spray on your grass? Or do you mow your own yard? If so, do you use a gas-powered mower? Did we know what kinds of chemicals our neighbors might be using on their lawns? Do you clean your own house or do you hire out this work? Either way, how attentive are you to the ingredients in your household cleaning products? How often do you vacuum your house? What are your carpets made of? How about your cupboards and bookshelves? Real wood or composite plywood? Do you dry-clean your clothes? Do your kids play with soft plastic toys? Do you use perfumes or other cosmetics? Fabric softeners? Scented candles?

Once again, it was quite a list. As Katherine and I answered, Donnay scribbled on his sheet. Here I was again, being asked to think hard about things to which I'd never given much thought. The setting was a bit less grim; we were sitting in my kitchen, after all, not in a preoperation waiting room. Nobody was sick. And, as far as I could tell, nothing in the house had caused me any harm.

Nonetheless, Donnay's list, like the list in the hospital, gave me pause. I'd lived in a lot of houses and apartments over the previous forty-five years, with little awareness of the potential toxins in my surroundings. What, over the long term, had I been marinating in? And, more important, now that I had children of my own, what kind of marinade had we unwittingly prepared in our own home? Neither of my children, so far, was experiencing any chronic health problems; no asthma, no headaches or "brain fogs," no learning issues at school. Neither had shown any signs of chemical sensitivity. But as Donnay was quick to point out, acute symptoms of chemical exposure are only one worry. At least as important are problems that crop up over a lifetime of exposure. Reducing a child's risk of asthma is one thing; reducing their exposure to carcinogens and endocrine disruptors is something else entirely.

Donnay put down his pen and took another sip of tea. "Okay. There are two possible effects of chemical exposure," he said. "One is immediate: 'I know this is happening and it is making me sick.' The other is the long-term buildup that you don't notice until you have a tumor or a terrible illness. These two things are not necessarily linked. People who aren't bothered by multiple chemical sensitivity can't have any confidence that they won't be affected twenty years down the line."

Cancer, it is true, kills more children under fifteen than any other disease. But when it comes to the long-term implications of chemical exposure, it's important to understand that we're talking about many more illnesses that cancer. We are also talking about insults to the endocrine system, one of the body's primary communications networks. Hormones are produced by endocrine tissues like those in the ovaries, testes, and pancreas, as well as the thyroid and pituitary glands. The hormones are secreted into the bloodstream, where they act as chemical messengers to organs in the rest of the body. In conjunction with the nervous system, hormones control everything from the body's growth and energy level to reproduction and responses to stress. During development, hormones within the womb determine which of a baby's genes will be expressed, and how often, over the course of its lifetime. "Nothing has been changed in the individual's genes, but if a particular note hasn't been punched into the music roll during development, it will remain forever mute," write Theo Colburn, Dianne Dumanoski, and John Peterson Myers in their groundbreaking book about hormone-disrupting chemicals, *Our Stolen Future*. "Genes may be the keyboard, but hormones present during development compose the tune."

Chemicals can mess with a body's music in a number of ways.

They can mimic estrogen, the female sex hormone, and androgen, the male sex hormone. They can imitate thyroid hormones, causing overstimulation. Alternatively, they can bind to a cell and block the cell's ability to interact with a normal hormone. Or they can interfere with the way hormones are made or controlled—for example, by blocking metabolism in the liver. Higher risks of breast cancer have been associated with exposure to estrogenic compounds, but also to depressed thyroid levels—both of which can occur when people are exposed to environmental chemicals.

"This is a degree of sensitivity that approaches the unfathomable," the authors of *Our Stolen Future* note. "If such exquisite sensitivity provides rich opportunities for varied offspring from the same genetic stock, this same characteristic also makes the system vulnerable to serious disruption if something interferes with normal hormone levels."

Once we had finished the questionnaire, Albert Donnay drained his cup of tea and stood up.

"Let's take a walk around the house," he said.

We began on the ground floor, where, it seemed, there was much that he was pleased to see. We have no synthetic wall-to-wall carpeting, which is often treated with toxic stain-resistant chemicals; our floors are made of oak and are covered here and there with area rugs made of wool. We avoid cleaning products that contain bleach and ammonia, and instead clean our floors, counters, and bathrooms either with plant-based commercial products or a homemade mixture of water, white vinegar, hydrogen peroxide, and lemon juice. We threw out our nonstick cookware a long time ago, and do our best to remove our shoes before we walk into the house.

Donnay liked all of this, especially the no-shoes-in-the-house rule, something he practices somewhat religiously in his own home. He also liked the fact that we generally don't dry-clean our clothes and when we do, we use an "organic" cleaner. For years, dry cleaners have used perchloroethylene as their primary solvent.

The problem is that a number of studies have shown a link between perc exposure and neurological problems, especially related to vision, in the people who work in the cleaners. (A study done in New York City showed that people who lived in apartments in the same buildings that housed dry cleaners were exposed to perc at nearly eight times the levels considered safe. Two of the study participants were nursing mothers; both had levels of perc in their breast milk from two to twelve times the safe levels. How big a concern is this? In New York City alone, 600 apartment buildings also house dry cleaners. That's about 170,000 people. And for the rest of us? It's safe to assume that three-quarters of us have measurable perc in our urine.)

Donnay was less impressed with our cheap vinyl window blinds, something I confess I had never thought about. Do you really want PVC plastic shedding from your windows, he asked me, smiling in that therapeutic "It's my job to reveal the obvious to you" kind of way.

Then we got to the bathrooms. Donnay glanced at the personal care products—baby lotions, soaps, shampoos—lined up on the shelves, and frowned. Virtually all of these products contained "fragrances," which typically are made with phthalate plasticizers—compounds similar to those found in plastic baby toys and water bottles. Evidence is mounting that these compounds mess with the body's endocrine system, and Donnay is especially concerned about putting these chemicals on warm, damp skin. A 2006 study published in the journal *Pediatrics* found that more than 80 percent of infants had been exposed to at least seven different phthalate metabolites, most likely from lotions, powders, and shampoos. This is particularly worrisome given the large percentage of an infant's skin that is typically covered with these products, and their developing metabolic systems.

In recent years, researchers have linked phthalate exposure to a wide swath of developmental, hormonal, and reproductive problems in both laboratory animals and humans. Of particular concern seems to be the damage phthalates can do to the DNA in adult sperm and to the reproductive hormones in fetuses and infants, especially males. Laboratory tests show that phthalates damage reproductive organs of both males and females, but the male organs—especially the testes, prostate, and seminal vesicles— appear most vulnerable, especially in the offspring of contaminated animals. Testicles atrophy or fail to descend. The opening of the penis appears at the base of the organ rather than at the tip (a problem known as "hypospadias"). Sperm counts drop. And on and on.

Public health experts worry that these hormonal shifts may also contribute to problems like breast cancer and uterine disease. In May 2002, a coalition of environmental and public health groups contracted with a national laboratory to test 72 name-brand, off-the-shelf beauty products for the presence of phthalates. They found these plasticizers in nearly three-quarters of the products, including 9 of 14 deodorants, all 17 fragrances, 6 of 7 hair gels, 4 of 7 mousses, 14 of 18 hairsprays, and 2 of 9 body lotions. The phthalates appeared in fractions ranging from traces to nearly 3 percent of product formulation. None of the products listed phthalates in its ingredients. Loopholes in federal law allow the $20-billion-a-year cosmetics industry to put unlimited amounts of phthalates into many personal care products with virtually no testing, monitoring of health effects, or labeling.

Do you really want all that stuff in here? Donnay asked.

Okay. So we would throw out most of what we had in our bathroom. Got it.

As we walked toward the basement stairs, I did my best to lighten the mood. You know, trot out a little toxic-chemical humor. I brought up a story I'd read about a mortician who'd shown up at his doctor's office with a pair of odd complaints: he had no libido, and he had grown breasts. Turns out the man regularly applied hormone-heavy embalming cream without wearing gloves. Once he changed his behavior, he returned to normal inside two years.

I'm sure Donnay would have smiled at this, but by the time I got to the end of the story we had already opened the door to the basement. One step down, and I could practically hear him cringing.

I had to confess: the underground air *did* have a sickly, rather sweet smell. Had I never noticed it before or had I just turned a blind nose to it?

I should say at the outset that my family does not have a garage—during the construction project, we had converted the attached garage to a den—so most of the extra stuff that has accumulated over the years has ended up in our cement-floor basement. Old rugs, boxes of books, camping gear—it's all underneath the house. All of it, Donnay was quick to point out, basically functions as a giant mold hotel—even in a basement that is comparatively dry, and in which two small dehumidifiers are running constantly. Get rid of all this stuff, Donnay said, or move it upstairs.

But what really got Donnay wrinkling his nose was off to the right, back behind an old pair of skis: a pile of fifteen paint cans, jugs of sealants and floor finishes, and a half dozen tubes of grout, caulk, and polyurethane. The sealants, he pointed out, were particularly worrisome: some tubes were covered with warnings that their contents were skin and eye irritants and could cause respiratory distress. Others warned of dangers to the neurological system.

Donnay gave me a look. For a family with a pretty clean house, this is a real problem, he said. What do you think happens to all the fumes coming out of these cans and tubes? They don't dissipate, especially when there are no open windows down here. Remember the garbage bag filled with cans of toxic chemicals? This basement is like a giant garbage bag. The fumes seep through walls and floorboards and—depending on how well a home's ductwork has been installed—circulate throughout the house. Ditto for the combustion gases leaking from the furnace and the water heater. Especially carbon monoxide. In winter, when your furnace is working overtime, your basement can be the warmest place in the house, and hot air—along with the host of toxic chemicals it carries—has a way of rising.

How bad can this be? I wanted to know. Haven't people been storing stuff in their basements as long as they've had basements?

Donnay looked at me and smiled. Yes, they have, he said, and that's a problem. You've got a degree in literature, right? Ever read any Victorian fiction?

A little, I said.

Well, you know all those crazy women in the attics, and all those ghosts roaming around ancestral homes? In the nineteenth century, Donnay said, an entire class of illnesses was given a name, neurasthenia, which doctors at first blamed on the stress of Victorian life. They were convinced that the fast pace of life—the printing press, the steam engine, the telegraph—had pushed people's stress to such levels that they were getting sick. Contemporary literary critics blame patriarchal oppression for such distress; why else would the victims almost always be women?

Donnay had a different theory. It wasn't stress—technological or patriarchal—that was causing so much illness. It was toxic chemicals, especially carbon monoxide from gas-fired lamps. Just look at the literature of the day, he says. Victorian stories and novels are full of perpetually complaining female patients who never seem to get better. Their big-city doctors do everything they can until, exasperated, they lock their despondent patients in a room, either in the family manse or a crumbling sanatorium. From there, it's a short trip to madness.

It's an interesting idea, I had to admit, not least because I'd never really thought about literature from a toxicologist's point of view. Donnay was happy to indulge me.

In Edgar Allan Poe's "The Fall of the House of Usher," the narrator is asked to visit one of literature's most famously creepy houses, to see a friend who is suffering from "acute bodily illness—of a mental disorder which oppressed him." The man lives in the mansion with his twin sister; the two never leave the house, and both—like the house itself—are suffering from an unsettling rot. When the sister dies, the brother and the narrator carry her coffin down into a dreary cellar that had been used to store gunpowder "or some other highly combustible substance." With one last look, they fit the lid on the coffin and screw it down tight. Over the next several nights, the brother—and the narrator—slowly lose their minds. They are oppressed by a "faintly luminous and distinctly visible gaseous exhalation that hung about and enshrouded the mansion." They notice strange sounds—footsteps, grating iron chains, a beating heart—coming from the basement. The narrator bolts from the house. Had they buried the sister alive? Were they crazy, or was it just the fumes?

In Charlotte Perkins Gilman's 1892 story "The Yellow Wallpaper," a physician moves his wife to a room in a creaky ancestral hall for the summer, to help her recover from a "nervous depression." Locked in solitary confinement, the wife, who narrates the story from her bed, slowly descends into paranoid madness. With little else to occupy her mind, she starts to obsess over the patterns in the room's wallpaper, the "great slanting waves of optic horror, like a lot of wallowing seaweeds in full chase." She blames her condition on a peculiar smell that has enveloped her senses like a toxic fog. "Now we have had a week of fog and rain, and whether the windows are open or not, the smell is here," she says. "It creeps all over the house. I find it hovering in the dining-room, skulking in the parlor, hiding in the hall, lying in wait for me on the stairs. It gets into my hair. Even when I go to ride, if I turn my head suddenly and surprise it—there is that smell! Such a peculiar odor, too! I have spent hours in trying to analyze it, to find what it smelled like. It is not bad—at first, and very gentle, but quite the subtlest, most enduring odor I ever met. In this damp weather it is awful, I wake up in the night and find it hanging over me. It used to disturb me at first. I thought seriously of burning the house—to reach the smell."

By the end of the story, the woman is crawling around her room on all fours, tearing at the wallpaper. After shouting for an ax and breaking through her door, her husband sees her, and faints.

Albert Donnay believes these characters might have been suffering from what he calls "haunted house syndrome," an illness that would be familiar to any Victorian—real or fictional—fated to spend a night in a gloomy mansion. "People will tell you that they saw ghosts coming in the window to strangle them, that they taste metal in their mouths, that they smell things, that they feel creepy-crawlies on their arm," Donnay said. "At the turn of the twentieth century, they were locking millions of people away in insane asylums, and it was true, some of these people were completely nuts. But this was also the time of the peak use of gas lamps. Freud was writing at the time that these symptoms were all due to sexual trauma or excessive masturbation. But these are quite physically real symptoms. It's just that people's senses were lying to them."

This would all be quaint, Donnay said, if the parallels to our own lives weren't so chilling; we are now exposed to chemical loads that would make the Victorians, a century and a half ago, seem pure by comparison. Chronic exposure to even low levels of things like pesticides, formaldehyde, or carbon monoxide have been shown to have a serious impact on our health. Indeed, it may be that we've become so comfortable with these products that we no longer see them—or their constituent chemicals—for what they are. It's like the dry cleaners, Donnay says. A customer walks in, smells the chemicals, and says to the person behind the counter, "How can you work in here?" The people who work in the cleaners—the very people most saturated by the chemicals in the room—don't even smell it.

"We all get neurologically habituated," Donnay said. "We need a certain amount of stimulus to fire our nerves. But nerve firing is modulated. If your exposure has been random and isolated, like the guy who goes binge drinking every three weeks, you don't habituate very well to alcohol. But the guy who drinks every day gets very habituated. We all have the ability to slide up and down the scale.

"Ironically, the people who are most sick often have a lower chemical body burden, because they've tried for years to reduce their exposures. They are careful about their food and water. They don't even go out much. But they are still more sensitive than people with higher body burdens. Once you are sensitized, even chemicals at low or nontoxic levels can provoke symptoms."

Albert Donnay believes that exposure to toxic household chemicals may contribute to an entire host of illnesses and disease, from attention deficit disorder and autism in children to chronic fatigue syndrome and multiple chemical sensitivity in adults. And even if people are not directly sickened by a single chemical compound, he says, constant exposure can lead to chemical or sensory sensitivities that can be utterly debilitating.

"Pesticides, formaldehyde, chemicals used during remodeling—these things can cause the olfactory nerve to become hypersensitive to any chemical stimulus," Donnay said. "The dose might not be toxic itself, but it can still cause a reaction. It's like caffeine, or alcohol, or cigarette smoke. Ex-users can become so hypersensitive that they react to much lower levels than nonusers.

"We all have a sensory radar system that lets us know when we are threatened by a novel exposure," he explained. "The sensitivity of that radar can be turned up or down by the pattern of our exposures. If we are healthy, we can tune it out. It's not that these levels of stimulus are necessarily toxic, it's that some people have become so hypersensitive to stimulus at *any* level. They just can't tune out the background noise."

Where a nonsufferer might smell perfume, or a lawn pesticide, or cigarette smoke, and react no more strongly than wrinkling their nose, someone with chemical sensitivity might kick into a full-blown physical and emotional panic.

"It's a tremendous burden," Donnay said. "These people go into 'fight or flight' even if what is present is only a benign level of exposure. But 'fight or flight' takes a huge toll physically and emotionally. It degrades your immune system, it degrades cardiac health." Equally troubling, of course, is that hypersensitivity can make people feel isolated, even paranoid. It's not just that they suffer; it's that they suffer from exposures that most people don't even notice.

In other words, we may no longer use nineteenth-century terms like "neurasthenia," but that doesn't mean that we aren't living in haunted houses. Or working in haunted offices. If the last century has done anything, it has introduced many more of us to many, many more chemicals. And it has, for some people, meant lives that can seem like an Edgar Allan Poe fever dream.

Just finding a time and place to meet with Debra proved unusually difficult. When I suggested we get together in April, she balked. Spring, she said, is when the state uses helicopters to spray for gypsy moths, which meant she would have to stay close to home for weeks at a time. Coffee shops were out of the question, because she is allergic to the chemicals used to clean them. When we finally settled on a date, in July, we had to find a room that had never been sprayed with pesticides. After some calling around, we found one in a public library outside Washington, D.C.

Debra had short black hair and blue eyes that can seem twinkling or melancholy, depending on the story she is telling. As we sat down, she began waxing nostalgic about a time in her life when she could walk down her block without fearing an attack brought on by dryer sheets.

"I remember before they put fragrances in fabric softeners, when you could go out and take a walk in your neighborhood," she said. "Now, no one in our group can take a walk because of the fragrance that gets vented out from the dryer exhaust. Even though we walk out on the sidewalk in front of a house, we smell it coming out of the back of the house."

When it comes to chemical exposure, Debra holds to what she calls the "barrel theory." A human body is like a barrel: it can hold, or tolerate, a certain volume of toxic chemicals. But when it reaches its limit—when it no longer detoxifies properly—it overflows. It is as if a safety valve has broken. The trouble is, most of us are exposed to so many chemicals, for so many years, that by the time we notice the valve is broken, the barrel is already brimming. The mystery may not be why people like Debra start to break down, but why many more of us do not.

"Once you pass the threshold, you never go back to normal," Debra said. "So many people are ignorant of this disease, it takes years after you have passed that threshold to find out what's wrong with you. So by the time you figure out you have multiple chemical sensitivity, it's way too late to get back to normal."

The compounds that cause Debra such trouble are collectively known as VOCs, or volatile organic compounds—essentially, any chemicals that "volatilize," or spread invisibly through the air. In my house, this would include obvious things like the cans of paints and thinners in the basement, but it would also include things like perfumes, deodorizers, rug cleaners, and detergents that can be found in almost every room.

"This illness completely changes your life," Debra said. "It changes what you eat. Not only do you have to avoid pesticides and artificial ingredients, you have to rotate your diet. With this illness, if you eat the same things every day, you will develop food allergies. It takes three days to totally process your food, so something you eat on Monday, you can't eat again until Thursday. I often tell people I'm allergic to certain foods, but everyone with MCS develops food intolerances and allergies.

"When we buy new clothes, we have to soak them in baking soda, or baking soda and vinegar. Some people use hydrogen peroxide. Even the dyes in cotton clothing can cause problems. Some people who are very sick can only wear all-white, organic clothing, with no dyes or colors or pesticides of any kind."

When Debra thinks back on how her barrel began to overflow, she tries—mostly in vain—to find the first drops in her childhood. She was born in Florida, but her family moved to southern Rhode Island when she was eighteen months old. She grew up near the Quonset naval station and lived there until she was seventeen. There was some industry around, but there was also a cleansing breeze from the ocean. Her father was a carpenter, so Debra was no stranger to construction sites. Back then, she said, builders used real wood, not plywood glued together with formaldehyde. Since it was the 1950s, everyone around Debra smoked cigarettes, which release not just carbon monoxide but dozens of other combustion chemicals as well. In college, where she studied languages, a third of her classmates smoked in class; some professors smoked like chimneys. Worse, the building had sealed windows; on exam days, the room was a noxious cloud.

Still, for years, Debra showed no signs of chemical sensitivity. Her barrel was filling but was not yet full. It wasn't until she went to work as a linguist for the federal government in the mid-1970s that her ability to detoxify was affected. Ten months into her career, she started spitting up blood. Doctors thought she had tuberculosis. The X-rays were negative.

Two months later, a woman in Debra's office threw up at her desk, apparently sickened by the paint fumes billowing in from a crew working down the hall. For the first time in her life, Debra started noticing the quality of the air around her. The windows in her building did not open. Air circulation was poor. Many of Debra's colleagues smoked on the job, and some were sickened by the poor air quality. Her building was routinely cleaned with products containing volatile organic compounds; walls and ceilings were painted with volatile paints. Mold and other microbial contamination were common; Debra noticed black streaks near the ceiling vents. Even the copy machine gave her headaches. Years later, a toxicologist's report would conclude that the chemical load in her office created "a high-risk environment for adverse health effects in susceptible individuals."

If many of the people in Debra's office could tolerate the cigarette smoke—and the cleaning solutions, and the paint, and the mold—without complaint, Debra could not. Just six months after she started work, she was diagnosed with asthma, allergies, and sinus infections. She started getting nosebleeds. She complained of intense fatigue. She installed a HEPA filter near her desk, but it did little to help her symptoms. She tried asking the painters to use low-VOC paints, or at least to work at night. Her requests were rejected as too expensive.

Around her first anniversary on the job, Debra took a month of unpaid leave to try to regain her health. Soon after her return, her vision started to blur. She was transferred to another office, on a different floor. People still smoked around her, perhaps a bit less, but the cleaning solutions and the carbonless copy paper were the same. The sinus infections persisted.

After five or six years of this, Debra was given workers' compensation for her sensitivity to tobacco smoke. But the smoke, if it was an initial cause of her distress, was just the beginning. Over the next decade, she became acutely sensitive to paint fumes, developing intense headaches, nausea, and intestinal cramping even when painters were working down the hall. By 1988, even the perfume worn by her coworkers made her ill.

By 1990, it wasn't just indoor air pollution that made Debra sick. One day in May, while walking across a parking lot, she looked up and saw a plane spraying Bt, a pesticide used to combat gypsy moths. The next day, she woke up acutely ill and, for the first time, began to feel like her memory was going as well. (She learned this the hard way: while backing her car out of her garage, she forgot to punch the button on her electronic door opener, and slammed her car into the garage door.) Bt spraying, she testified before a state pesticide board, has added short-term memory loss to the headaches and "brain fog" she had suffered for years. For nearly twenty years, Debra has had to leave the state for three weeks whenever the helicopters come out to spray. She used to escape to visit relatives in North Carolina, but now gypsy moths—and the people spraying for them—have spread south as well.

A month after the garage door incident, Debra's troubles began to snowball. She was diagnosed with a pair of tumors in her thyroid. (They were benign.) Two years later, she was diagnosed with adrenal problems, which were causing her night sweats and chronic fatigue. She got muscle spasms from the shampoo used to clean the office carpet. She got sick from the formaldehyde in the glue in the office furniture. Fumes from paint and perfume and printer ink continued to be a curse. The smell of microwave popcorn became intolerable, as did the grease used on the office escalator. In 1998, her thyroid tumor began growing again. Her short-term memory loss recurred; after a week of exposure to a colleague's perfume, Debra—a linguist—could not remember how to spell her own name. The mold in her office, resulting from a leaky roof, only magnified her symptoms.

After twenty-five years, at the age of forty-seven, Debra decided she could no longer work. By then, she had finally found her way into the care of a doctor familiar with multiple chemical sensitivity, or MCS. He attributed her conditions to "multiple acute exposures and continuous low-dose exposures" to both toxic chemicals and biotoxins like mold and bacteria. A neurotox-icologist concluded that her health effects "are likely to be irreversible."

The mechanisms for MCS are mysterious. When a sensitized person is exposed to a certain airborne chemical—a pesticide, say, or a perfume—the chemical passes through their nose to the centers in the brain that control the way their body operates. "Those centers in your brain get hammered and they start misfiring," Debra says. Your chest tightens. Your breathing speeds up. You can't digest your food. You get dizzy. You forget things, like how to add numbers in your head. You become clumsy. You get depressed. Synthetic fragrances can give Debra horrible headaches that can last three days. A friend of hers had it far worse. When she walked by someone wearing perfume, she would lose her balance and collapse.

"Some products will say they are 'unscented,' but if you look on the list of ingredients it will have 'masking fragrances,' which still give off VOCs," Debra says. "People will say, 'Gee, how can she be reacting to this? I can't smell it. She must be crazy.'" Indeed, like Albert Donnay, Debra has a theory about all those women fainting their way through nineteenth-century novels: they had MCS. Their drawing room swoons weren't caused by a gentleman's gaze. They were caused by his perfume. Or their own.

When Debra began looking for a house, she had to find one that was at least five years old, just to make sure the formaldehyde in the plywood and the VOCs from the paint had had enough time to off-gas. Even house hunting was problematic, since people trying to sell their homes tend to use of lot of air fresheners, paints, cleaning chemicals, and lawn-care pesticides. Imagine asking a real estate agent to show you a house that hasn't been "treated" this way, and you'll see what people like Debra are up against.

So what causes someone to become hypersensitive to synthetic chemicals? There are theories: that chronic exposure to low levels of chemicals corrodes the body's nervous system; that it suppresses the immune system; that chemicals can lead to "limbic kindling," a triggering of the pathways between the olfactory nerve in the nose and the limbic system in the brain, which governs everything from sleep and mood to appetite and impulses like aggression. In short, no one knows for sure.

But there does seem to be a consensus on a few key points. The condition is chronic and can be brought on by low levels of exposure to chemically unrelated substances. Symptoms involve multiple organ systems, including respiratory, musculoskeletal, endocrinological, immunological, and others. Symptoms improve when chemicals are removed.

Yet the broader medical community remains at odds—and even downright skeptical—about the causes of MCS, with perplexing implications for people suffering from it. Some physicians insist that MCS is a purely psychological condition, and try to convince patients to overcome their fear of exposure by exposing themselves to *more* chemicals. Other physicians, those who consider MCS a physiological syndrome, advise the opposite: they tell patients to avoid shopping malls and other crowded places where they are more likely to encounter perfumes or detergents that can set their symptoms aflame.

It is true that MCS leaves psychological scars. Worse yet, people who get blindsided by subtle odors—people whose brain chemistry has been adversely affected by neurotoxins—can develop bouts of depression, anxiety, even agoraphobia; in her book *Multiple Chemical Sensitivity: A Survival Guide*, Pamela Reed Gibson writes of people who were so affected by their carpets and plywood they began living in cars. Or tents. One poor soul lived in a horse trailer for a year.

"Imagine a situation where you have a chronic illness where there is no known treatment," Debra told me. "You spend all your money trying to get well, trying to take the edge off the headaches. You're broke, and you can't find safe housing. If I buy a new car, I have to keep my old car for a month or two and run an ozonator in the new car to get the 'new car' smell out. Every aspect of our lives is controlled by this. What you eat. What you wear. Where you go. When I go to church, I have to go early and sit in a corner to be away from everybody."

The trouble for people like Debra is that these chemicals can pass through the body so quickly that even were she to be tested immediately upon exposure, the chemicals might not show up in her blood. But just because a body has voided itself of an irritant does not relieve a person from suffering its effects.

"Everything you do is dictated by your illness," Debra continued. "You have to be constantly vigilant to avoid or lessen your exposures to the greatest degree possible. The best thing we know to do with this illness is avoidance. If we take medicine, we usually get all the side effects, and the side effects greatly outweigh the benefits. Doctors get frustrated with us because they can't sit down and write a prescription. We have to start very gradually, with much smaller doses, and even then we may never tolerate it. It's very individual, from person to person, what will help them and what will not, what we react to and what we don't. And that has stymied the medical community for a long time."

You have to wonder: With so many chemicals wafting around us, throughout our lives, who wouldn't become distraught, especially if one doctor tells you to avoid crowds and another tells you to get over your fears and wade right in? Especially when doctors' offices themselves (to say nothing of hospitals) are saturated in the very chemicals people with MCS are sickened by: plastic tubing, vinyl blinds, endless gallons of cleaners and perfumes and pesticides. And especially when people with such an ambiguous disease are dismissed as malingerers or, at best, neurotic.

"We are a culture that wants desperately to believe that a positive attitude will make everything right," Pamela Reed Gibson writes. "This attitude can be tricky, however, when you are nose deep in a neurotoxic exposure. Because neurotoxins alter brain chemistry, simply trying to maintain a positive attitude will not be enough to stay healthy."

In the mid-1990s, Albert Donnay helped found MCS Referral and Resources, a national nonprofit organization that works with physicians and researchers to document cases of multiple chemical sensitivity and to look for causes and cures. Donnay gets hundreds of calls a year from people claiming they've been poisoned by their homes, their lawns, or their offices. "These people will say, 'I'm your canary in the coal mine,' and indeed they are," Donnay told me. "Their alarm bells are always going off, even if yours aren't."

All of which made me interested to hear what Donnay would say once he got up off our kitchen floor. We had climbed up out of our basement, and Donnay was on his hands and knees, sniffing under a Persian rug that had been in my family for at least fifty years. C'mon, I thought, this thing is made of wool. How bad could it be?

Donnay looked up, already one step ahead of me. It's not just what the rug is made of, he said. It's what the rug collects, how long it holds on to it, and how much time we spend breathing it in. Donnay is most concerned about compounds that are emitted by a long list of home furnishings made with toxic chemicals. Rugs and mattresses (made with toxic flame retardants), cabinets and wallboard (glued together with formaldehyde), interior paints (full of volatile organic compounds), even vinyl window coverings and shower curtains (made with plastics like phthalates and PVC)—all of these contribute to a level of interior air contamination that can lead to both short- and long-term health problems. Off-gassing follows a parabolic curve, Donnay said. When you first pull a mattress from its plastic sheath, or first lay down wall-to-wall carpeting, the chemical releases are very high. They decrease quickly in the first few weeks, but may continue at low levels for months or even years.

But it's not just air pollution that causes problems. The same products that produce toxic gasses also gradually break down into dust particles that can accumulate in a home and stick around for years. A person's home, like a person's skin, may provide a thin membrane that protects us from environmental toxins. But how much protection can the home, or the skin for that matter, offer us when our walls and our skins themselves are so saturated? The plywood sheath that wraps most of our homes is typically glued together with adhesives containing formaldehyde. (Nearly seventy thousand relocated victims of Hurricane Katrina learned this when their FEMA trailers were found to contain formaldehyde at seventy-five times the safety threshold.)

Donnay recalled one woman, in Philadelphia, who called him to complain of a constant pain in her chest. Her lungs were burning, she said, and her brain seemed to be in perpetual fog.

The woman, it turned out, had become so sensitive to carbon monoxide, pesticides, and other chemicals in her home that finding a safe place to live had become next to impossible. Once, after she returned from a vacation to find mold from a water leak, she had to throw away her furniture. She can't have gas appliances, she can't have fresh paint on the walls, she can't spray her home or her garden with pesticides. Worst of all, she can't distinguish one contaminant from another.

When she called Donnay, she thought her problems were being caused by an allergy to her dog.

"The moment I walked in the door I smelled paint," Donnay said. "She had a terrible basement. It was full of all kinds of undecipherable odors. She had taken out all the rugs except the wall-to-wall on the stairs. We got down on our knees and sniffed it. The fabric was really off-gassing." Not only could Donnay smell the chemical stain resisters and flame retardants that had been infused into the carpeting, he could smell other things that the carpet had absorbed over the years: mold, soot from the fireplace, organophosphate pesticides—all of which had settled into the carpet "like a sink."

And so it is for many of us. Despite our sense of ourselves as a nation of rugged pioneers, Americans have in essence become a nation of great indoorsmen. On average, we spend approximately 90 percent of our time indoors—more during the winter months, and most of this at home. And dust, of course, is not only a by-product of construction. A 2003 study of private homes found dozens of toxic chemicals hiding in indoor dust, including strikingly intact quantities of DDT, which has been banned since 1972, and a flame retardant that has been banned since 1977.

The EPA has warned that indoor air pollution can be two to ten times worse than outdoor pollution. In Baltimore, one in five children suffer from asthma; public health experts blame everything from toxic flame retardants in carpets and indoor bug sprays to lead dust and the chemicals from (of all things) air fresheners. Pesticides drifting into homes from farm fields can concentrate in carpets at 200 times the rate in outdoor soil. The allergen level of sealed buildings can be 200 times that of older buildings. "Anything that goes into the air, anything you spray, will interact with the dust in the air and settle down," Donnay said. "On smooth surfaces you can wipe it clean. But on rugs it's like planting the dust in a garden. Even a vacuum won't get it out. Just like you don't want to wear clothes that you just brush clean, you don't want to live with rugs that are just vacuumed. In the old days, people took rugs outside and beat them with a stick. You should wash them at least once a year, even if it means taking them down to the commercial Laundromat and throwing them into one of those machines that take monster loads."

Household dust is a real concern, Donnay said, because it can accumulate not just in rugs but in the forced-air vents that heat and cool so many homes. Compounding the problem, some companies use a "wet" cleaning system that includes the use of scented cleaning products—precisely the kind of compounds you're trying to get rid of in the first place.

"These chemicals enter your body, and not all of them will get out again," Donnay said. "You breathe some out, but you metabolize some as well. The body is great chemical factory, and it's running all the time. It doesn't benefit you to add to the ingredients in that chemical factory. Whether you add it to your skin or breathe it, any chronic lifetime exposure is going to increase your body burden, and therefore increase your potential pathways for disease.

"Obviously, we use these things in combinations," he noted,

"but that's where we've reached a dead end. We don't have the tools, the money, or the time to understand these combinations of chemicals. We are the experiment, and our health is the outcome."

In 2003, a team of researchers led by Ruthann Rudel, of the Silent Spring Institute, swiped samples from 120 private homes and discovered dozens of toxic chemicals lingering in household dust. Not just flame retardants but phthalates and disinfectants and even pesticides. Among their striking discoveries were "outdoor" compounds like DDT, heptachlor, and chlordane, which have been banned in this country for years and, in some cases, decades. All told, the researchers discovered 52 compounds lingering in the air and 66 in household dust, including—in both categories—some two dozen pesticides. Fifteen pesticides were discovered at levels exceeding government safety standards; the risks for another 28 could not be adequately assessed because, for them, no health standards had ever been set.

What the study revealed was not just how ubiquitous these compounds can be but how long-lasting. DDT has not been sold domestically for more than thirty years, and there it was, in people's houses. Setting aside the question of how an outdoor pesticide makes its way inside a house, who knows how long it lasts when it is not exposed to the damaging effects of sun and weather. The breakdown of pesticides, once inside the home, is "negligible," the report said.

And the human inhabitants of polluted homes are hardly the only ones vulnerable to toxic contamination. If you walk across a carpet treated with flame retardants or stain-resistant chemicals, you may kick up some dust, or some of it may stick to the bottom of your shoes. But your dog or your cat spends hours every day rolling around on the floor, or sleeping, or—how best to say it?—licking themselves. The synthetic chemicals in the dust in the rug end up on your pet's tongue. The EPA has been exploring a possible connection between flame retardants and feline hyperthyroidism; cats, researchers have found, have contaminant levels up to 100 times that of humans. Richard Wiles and his colleagues at the Environmental Working Group took samples from a few dozen healthy dogs and cats in Mechanicsville, Virginia. The animals had levels of phthalates that were lower than they typically are in people. That's the good news. The bad news is that dogs had more than twice the levels of toxic stain-resisting chemicals than people. Cats had 5 times the mercury. And the levels of flame retardants were 23 times higher. "They're picking up the same chemicals that we're exposed to," Wiles said. "They have shorter life spans and they develop diseases more quickly, and so they may be providing some insight into human health problems from these same contaminants in our homes."

· Before Albert Donnay left, I invited him back into our kitchen for a final question. I wanted to know how likely our house was to become haunted by carbon monoxide. We don't have an attached garage, but we do have a fireplace, our furnace burns natural gas, and we have a gas range in the kitchen. Was anyone in our family likely to go crazy and try to bury me alive?

Although our bodies constantly produce carbon monoxide as a low-level metabolic by-product, adding external sources "can seriously corrupt our body's ability to handle the flow of CO," Donnay said. The carbon monoxide that doesn't bind to hemoglobin floats around the body, looking for other places to attach itself. It can end up in muscle cells, where, he noted, it can lead to fibromyalgia; it can find its way into the body's mitochondria, where it can inhibit metabolism and contribute to chronic fatigue. It can trigger acute sensory sensitivity to odors, light, sound, or touch—all symptomatic of disorders like autism and Asperger's syndrome.

My wife, Katherine, had installed a carbon monoxide detector right next to the stove. Wasn't that a smart idea? I asked Donnay.

Well, for starters, Donnay said, your range hood only covers about two-thirds of the stove top, and I'm guessing you don't even turn it on unless you're cooking a full meal. Am I right?

I gave him a sheepish smile and nodded. I'd certainly never turned the fan on when I'd made a pot of coffee.

This jibed with Donnay's own experience. Most people think

the vent is intended to shed the smell of burning food. They don't even bother to turn on their vents unless they smell smoke.

Donnay took out his own professional-grade CO monitor. The EPA's standard limit for outdoor carbon monoxide exposure is 9 parts per million, he said; now watch this. He turned on one burner at a time, and waited. One hit a peak of 125 parts per million before settling back to 80. Others peaked in the mid-50s or the mid-70s and settled in the mid-20s.

Our own plug-in CO monitor, mounted inches from the burners, remained remarkably consistent. It never moved away from 0.

Even six feet from the stove, above our kitchen table, CO levels hovered at 9 or 10—the EPA's safety limit. And these were readings after running the burners for just a couple of minutes. Imagine, Donnay said, how much carbon monoxide you'd have after cooking a six-hour Thanksgiving dinner? As for the so-called self-cleaning feature on the oven? All that does is turn food residue to ash—carbonizing it—and create carbon monoxide levels of several thousand parts per million.

So what was the point of using a commercially available CO detector, if it registered 0 even when CO levels were in the hundreds of parts per million? When Donnay checked a monitor we had placed in our kids' upstairs bathroom, he found it had at some point registered a maximum level of 13—yet its display had never wavered from 0.

Donnay pulled our store-bought monitor from the wall and held it in his hand. "This thing isn't even allowed to alarm until the CO gets over seventy parts per million for one to four hours," he said. "The Consumer Product Safety Commission doesn't even want you to know about CO until the level is over thirty. Women can have babies with birth defects and low birth weight when average background CO levels increase only five to six parts per million. Imagine if smoke detectors had to wait four hours to warn you of a fire!

"These decisions were made strictly for convenience of fire marshals and gas companies," he went on. "The manufacturers say, 'Do not put CO monitors in the garage or kitchen or furnace room' because they don't want to overburden gas companies and fire departments who must respond if they get called about a CO alarm. If they find low levels in the house, they say, 'This detector must be defective.' Even if they find high levels, they only inspect gas appliances, and never the cars in the attached garage. They commonly say, 'We can't find the source of all this CO.' But studies have shown high CO levels in the house several hours after the car has left the attached garage. It takes that long for CO to seep into the house. And how are you even supposed to test these CO detectors? All of these, when you hit the test button, you're only testing the battery and the alarm. There's no way to know if the sensor is even working. It's very hard to set these things off. I've found they can be one hundred percent off—if the actual CO level is one hundred, the monitor could display two hundred, or it could display zero."

Donnay speculates that the epidemics of asthma and autism that have arisen in recent decades are due at least in part to two common sources of CO poisoning: the use of poorly ventilated gas ovens and stoves (half of American kitchens have gas ovens) and the widespread construction of homes with attached garages.

"We brought our cars into our houses, took the flues off our ovens and replaced them with dummy hoods that blow the CO right back into our faces," Donnay said. Especially when the weather (and thus car engines and the catalytic converters) are cold, starting cars inside garages produces a lot of carbon monoxide. Even though the car's exhaust pipe is typically facing out, once the CO hits the cold air outside the garage (Donnay calls this air "the glass wall"), most of it rushes straight back into the garage. Once the car leaves, and the garage door drops down, the CO remains locked inside, where it inevitably seeps into the house, resulting in levels high enough to trigger CO alarms—provided the alarms actually work.

Donnay listed other sources of carbon monoxide poisoning, noting especially gas and charcoal grills and poorly ventilated indoor fireplaces. Time was, people put their coals in a coal bucket and took them outside, he said. In contemporary fireplaces, as the fire dies down, the force of the updraft gradually dimishes. Long after you've gone to bed, cold air rushes down the chimney, fans the remaining embers, and blows combustion gases throughout the house. (The solution? Install a ventilated "insert" box with tightly closing glass doors that will allow you to enjoy a fire without allowing the combustion gases to reverse course.)

Gas-powered lawn tools are another source of acute carbon monoxide poisoning, Donnay said. "Pick any two-stroke engine—mowers, leaf blowers, weed trimmers—and you're getting tens of thousands of parts per million of carbon monoxide exhaust. There's no chimney on these appliances, and no catalytic converter to reduce the CO. You're either wearing it on your back or pushing it right in front of you. You're basically walking in a CO cloud as long as you're using that appliance."

As someone well versed in the politics of environmental health, Albert Donnay acknowledges that his ideas about carbon monoxide are not exactly in the mainstream of medical science. He is also aware that conditions like multiple chemical sensitivity offer skeptics an opportunity to disregard any complaints about chemical exposure. That a few poor souls are suffering from acute sensitivity is unfortunate, this thinking goes, but that's their problem, and not a problem with chemicals themselves. If the majority of us seem to get along just fine, shouldn't we leave well enough alone?

The weakness of this argument, Donnay says, lies in the difference between a person's chemical sensitivity and a person's chemical body burden. It is true that people with MCS suffer in ways that most people do not. Since most of us are not chemically sensitive, since encounters with pesticides or cleaning fluids or paints do not send us reeling, we are not inclined to see our chemical exposures as threatening. But the thing about synthetic chemicals is that they do not just cause immediate problems. Far more worrisome are the problems they cause as they accumulate over time. It's one thing to realize that you are sensitive to your husband's deodorant. You just have him throw it away. It's quite another thing to realize that you have years of toxic flame retardants built up in your breast tissue.

"A drug addict or a chain smoker or an alcoholic may not be cognizant of their body burden, either, but they are still likely to die young because of it," Donnay says. "We all are accumulating these chemicals, but what effects they may have on us, we still don't really know."

THREE The Big Box Store

Now that we'd had our eyes opened a bit, and had absorbed Albert Donnay's assessment of our home, Katherine and I decided to explore the provenance of the myriad toxins streaming into our lives. To follow the river, as it were, to its very source. Not to the wellspring, exactly—we weren't going to go undercover in any Chinese factories or look inside the cauldrons of a chemical multinational—but to the grand municipal reservoir, where these chemicals accumulate in unimaginable volume before being piped in a steady stream into our homes. Rather than run around from hardware stores and pet shops to furniture outlets and clothing boutiques, we decided to shop where millions of Americans shop every week, for everything from mattresses and toys to cosmetics and even food. It would be a kind of safari, if you will, full of mysterious landscapes and more than a few hidden dangers. We would walk slowly and deliberately through a big box store. We would turn over a few rocks, and see what we could see.

Let me begin by admitting that I have an intense, even neurotic dislike of shopping. When I need something—a tube of toothpaste, a can of paint, a new pair of pants—I zip into a store, grab what I'm after, and zip out. I don't browse, or linger, or paw through the racks. If I see a couple of brands side by side, I'm likely to pick the one that is cheaper, and I'm not likely to think very hard about why the price might be so low. I'm cheap that way.

Which, I suppose, makes me a pretty typical consumer in a typical big box store. And big box stores attract a lot of typical customers. In 2009, the top ten big box chains, led by Walmart, Costco, and Target, sold merchandise worth more than $1 trillion.

It's worth saying right up front that there's very little for sale in a big box store that you can't find in dozens of mom-and-pop stores. It's just that in many places, big box stores have put all this stuff under one roof (and in many cases have also put all of those other stores out of business). It's also worth remembering that big box stores, at least where I live, come in clumps. In a typical mall around Baltimore, you are likely to find a Walmart sharing one wall with a Target, another with a Michaels Arts and Crafts, and a third with a Toys "R" Us. Though their marketing strategies may vary, these stores share a great many products, which in turn share a great many toxic chemicals.

The trouble is, other than price tags, the products for sale in most of these places provide very little useful information. Although virtually everything is stamped "Made in China," there is almost nothing else to go on. Huge swaths of products, from plastic toys to upholstered furniture, report nothing about what went into making them—not the materials, not the circumstances under which they were manufactured, not whether or not they were "inspected," nothing. For the products that do list a few ingredients— paints, for example, or caulks, or some laundry soaps—the labels are typically inscrutable, Latinate, and printed in fonts that seem purposefully difficult to read. It's like the information was created to *dissuade* you from asking too many questions.

In this way, big box stores are not unlike fast food restaurants: you walk in looking for a very inexpensive product, and that is what you get. Where the thing was made, and with what ingredients, is simply not part of the equation. It's not what you came looking for, and you're not going to find it.

Which doesn't mean that there's not more to know. Peel back the curtain on fast food, and you'll find a cheap hamburger whose provenance—the industrial food machine—has successfully hidden a full range of social, environmental, and public health costs. Look a little closer at the products in a big box store, and you'll find an equally full range of compromises: a huge petrochemical industry; a shocking lack of government oversight; an endless conveyor belt of products moving into this country from factories somewhere in China. There is a lot to look for in a big box store. It's just really hard to know where to find it.

After parking our car out in the vast acreage of our suburban Baltimore megamall, Katherine and I made our way through the automatic doors and were immediately struck—we were looking with new eyes, don't forget—by just how perfectly denatured the big box environment was. There were no windows, so once inside the doors we lost all sense of any life going on outside. There were no clocks, so we quickly lost track of time. The store was so big we couldn't even see its walls; wandering the aisles, I began to feel as though I were adrift in a borderless, rolling sea of merchandise. The space seemed disorienting by design.

And it wasn't just the visual stimulation that overwhelmed us. It was also the smell. The moment Katherine and I walked through the door, we were hit with a blast of fragrance. Or perhaps I should say "fragrance." At first, the place smelled vaguely familiar, like the doorway of a shopping mall candle store or a gas station bathroom: there was a powerful bouquet of vanilla and apple cinnamon, coupled with faint notes of . . . lemon? At the back of the palate, I thought I could detect a few notes of burning rubber, with a subtle chlorine nose. But the finish? What *was* that? I knew I'd smelled it before, just not in such vivid concentration. What was that? Was that "lavender"?

Katherine and I looked at each other. Wow, she said. That's very strong. I wonder where it's coming from? We looked around.

To our left was the wing of the store devoted to personal care products. Was that it? Was this olfactory cloud coming from a brigade of shampoo bottles? Or was it drifting in from the laundry detergents? Was it over there, to the right, in among the scented candles and potpourri?

The truth was, it was impossible to tell. Almost every single product in the store, from the paints and caulks to the cosmetics and the baby bottles, had been manufactured, one way or another, from a mix of synthetic chemicals. The smells were all so ambient, so pervasive, that discovering their provenance was clearly going to take some work.

It was about this time that an image popped into my head: Albert Donnay standing atop a giant crane, wrapping the big box store in an enormous plastic bag. For the rest of my visit, I couldn't shake the image of this bag, sealed for a month, and then opened like a giant Pandora's balloon.

HARDWARE

Katherine and I decided to look first in the aisles featuring products that had raised so many red flags in our basement. If such a sickly odor in our home had been caused by just a dozen cans of paint, we figured we ought to take a look—or a smell, if you will— at the hardware department.

When it comes to toxic chemicals, the hardware section in a big box store is a red zone. On shelf after shelf there were lacquers and paints, strippers and thinners, grease and stain removers—the same stuff we had fumigating our home from below. These products were, for the most part, canned or shrink-wrapped. But as I held them in my hands I couldn't help thinking about all the toxic genies waiting to be released from their bottles.

In the first aisle, Katherine and I waded into what I came to think of as the Pond of Caulk. This was a place I imagined spending some money; the grout in our bathroom had begun to chip and crack, and I needed to figure out a way to seal the fissures between the tiles. I picked up a tube and looked. It said it contained ethylene glycol, gamma-aminopropyltriethoxysilane, formaldehyde, ammonia, and acetaldehyde. Some of these things, the packaging said, "may cause kidney, cardiovascular and liver effects." I put it down, gently. Another label said the product contained "phthalate esters"; this, to be honest, deserved some kind of award. Of the thousands of products in the store that contained phthalates, this was the only one that I saw that actually acknowledged it.

There were sealants guaranteed (on the front) to last fifty years and warning (on the back) that the compound can be fatal if swallowed and "contains chemicals known to the state of California to cause cancer and birth defects or other reproductive harm." In microscopic print, there was an advisory about how to dispose of the chemical, along with this bit of news: "Reports have associated repeated and prolonged occupational exposure to solvents with permanent brain and nervous system damage."

A number of these products recommended that they be applied with rubber gloves and that masks be worn. I found myself thinking back on all the home repair projects I had done over the years: all the painting, floor sanding, and caulking; all the fiberglass I had installed in my attic and the sealants I had put on my porch floor. I thought about the epoxies and resins I had used to fiberglass three canoes I had built out of strips of western red cedar; despite the warnings on the resins' packages, I had painted it on with bare hands. And since I'd built them outside, I'd never worn a mask.

I'm sure there are some people who wear gloves when they re-caulk their bathtubs. But to be honest, I'd always thought a wet finger was the best way to apply the stuff. And I bet there are plenty of people who, when they see a stray swab of white gunk, simply wipe it up with a bare finger. Ditto for parents painting their kid's bedroom, or staining a bookcase in the garage, or rolling out a pile of pink insulation in their attic. And then, of course, once the stuff has been applied, we all have to live with it as it breaks down, becomes dust, and begins filling our homes.

Katherine picked up a tube of silicone. This might be better than caulk, she said. Isn't this the stuff they put in breast implants? How bad can it be?

I don't claim to have the first idea about what goes into a breast implant. But finding out what was in this particular tube of silicone caulk was tricky. It came with a multipage label, with the ingredients listed inside. When Katherine read it, she found that the caulk "may cause kidney, cardiovascular and respiratory problems." If she wanted more information, she was offered a phone number.

"Just the fact that you have to go through all these steps to figure out if something is safe makes me nuts," Katherine said. "And look at this." She showed me the label. "Manufacturer can't be held liable for damages in excess of purchase price," it said. What damages were we talking about, I wondered? Damages to your bathtub? Or damages to your body? And how much damage would $1.97 cover, anyway?

As I was looking these things over—I am not making this up—I saw a man with a woman, apparently his mother, examining a selection of sealants. They seemed to be trying to find a way to keep the winter air from slipping through their leaky window frames. "This will get rid of it," the man said, and he and his mother took the package to the checkout counter. I couldn't help but wonder about the quality of the air in his soon-to-be-sealed home, once he got "rid" of the clean air and layered it over with a thin barrier of toxic chemicals.

One thing that distinguished the hardware section was the presence of warnings on most labels. Cans of thinners and strippers were marked with "may be fatal or cause blindness." There were deck strippers that cause "severe burns to eyes and mucous membranes" and that contain chemicals "known to the state of California to cause cancer." (This was the second time I had seen this warning. What was it about California? Did these products cause cancer in California and not in other states?) Again, it was hard to tell which chemicals they were talking about, though one package did list sodium hydroxide as a possible suspect. (Another irony: these products were being sold right below the humidifiers. I remembered something Albert Donnay had said about the toxins in bath products, which enter the body all the faster when the pores of a person's newly bathed skin are wet, warm, and wide open.)

The spray-on "popcorn texture" for ceilings said it contained an ingredient (though it didn't say which one) "known to cause cancer and birth defects or other reproductive harm" (again, in California, anyway). Other ingredients in this product, I was informed, may affect the brain or the nervous system.

The popcorn ceiling spray was, for me, a kind of (unsavory) Proustian madeleine: I immediately thought back fifteen years, to when Katherine and I had briefly found ourselves renting a house in suburban Pennsylvania. We had chosen the house for one main reason: the owner had allowed us to live there with our two dogs. As we toured the house, the landlord had boasted about the popcorn (or, in his words, "Hollywood") ceilings, including one right above our bed. (He had also boasted about how he had outfitted much of the house with floor-to-ceiling mirrors, but that's another story.) So desperate were we to find a place that accepted dogs, Katherine and I signed the lease, and we lived there for a year. Now, standing in the hardware section, I wondered: How much of that Hollywood ceiling had ended up in my lungs? And once in there, what had it done?

The hardware aisles were also full of every kind of aerosol spray, each with its attendant warnings about chemicals and their potential effects. Polyurethane (ketones, butane; brain and nervous system damage). Lacquer (acetates, ketones; ditto). Polycrylic finish (dimethyl ether, 2-butoxyethanol; ditto). Peekaboo Blue spray paint (acetone, toluene, aliphatic hydrocarbons; ditto). Rust-Oleum (toluene, acetone, xylene; ditto).

I had always pooh-poohed these warnings, but as Albert Donnay had reminded me, these side effects are symptoms of neurological distress. How many of these exposures does it take to add up to a neurological problem?

The truth was, after a half hour spent wandering around this big box store, I was starting to get a nasty headache. Whether this was from chemicals leaking out of individual products, or from the cumulative "fragrances," or from straining to read all those microscopic labels, I can't say for sure. I will also not speculate on whether this headache qualified as "neurological distress." As I say, I'm not a particularly joyful shopper on a good day.

We kept looking, pushing our cart to an aisle given over to what I came to think of as the Arsenal. Here, it seemed, you could buy weapons to eradicate every living organism smaller than your thumb. There were aerosol ant killers, roach and termite sprays, flea powders, atmospheric "foggers." "Kills on Contact," these products brag. "Once and Done!" "Kills 180+ types of insects." "Kills Bugs Inside and Out." These products are intended to be shaken or sprayed over the width and breadth of indoor carpeting, and warn that you should wash your hands and clothes after use. They say nothing about the consequences of breathing these poisons for weeks or months; as is true for so many kinds of dust, indoor pesticides tend to linger in carpeting for a very long time without breaking down. Ditto for the aerosol flea "bombs," meant to be set up in a sealed room until the entire space is fogged with pesticides. Users are advised to call poison control if the stuff got swallowed, but there are no ingredients listed for 96 percent of the contents. Here's what is listed, labeled as "active ingredients": linalool, piperonyl, pyrethrins. The last of these represented some of the pesticides the CDC had recently discovered "in much of the U.S. population."

These chemicals are designed to kill insects living outside the bodies of dogs and cats. But as Albert Donnay pointed out, aerosolized chemicals also enter lungs. And once powders settle into carpeting, they end up on animal fur, adult feet, and infants' hands—from whence, whether through skin absorption or inadvertent oral consumption, they also enter our bodies. These compounds include organophosphate and carbamate pesticides, chemicals designed to damage a bug's nervous system.

And in people? While some effects on human health—such as vomiting or tremors—may be immediately apparent, other impacts are more pernicious. Childhood exposure to pesticides—

both in the womb and during the first years after birth—has been linked to an increased risk of cancer and to increased risks of injury to both the developing brain and the nervous system. Pound for pound, one study reports, children are at far greater risk than adults "because they live and play close to the floor, breathe close to the ground and constantly put their fingers into their mouths."

And finally we came to paint. Now, there may be some people who wouldn't know a sealant from a caulk. I myself remain confused about the difference between lacquer and polyurethane. But everyone knows about paint. We coat our houses with it, inside and out. We spend a great deal of time thinking about texture (forget gloss or semigloss; now companies can set you up with paints that glow with reflected light, or shimmer like sprayed-on metal, or look for all the world like suede). We think about fashion (Sherwin-Williams now sells a paint called Recycled Glass.) Most of all, we think about color (Ralph Lauren boasts that it carries over sixty shades of *white*).

What we don't think about, at least not too deeply, is what goes into the stuff we spread all over our walls and ceilings. Or what happens when we breathe it. Or when it breaks down and ends up in the dust in our rugs, which ends up, over time, in our skin, and our lungs, and our food. On the front of one can, I read: "Recommended for family rooms, children's rooms and hallways." On the back: Contains crystalline silica. Long-term effects: abrading or sanding "may release crystalline silica which has been shown to cause lung damage and cancer."

Next to this was a can of interior semigloss paint that offered a block-lettered public health warning—not about itself, but about the old paint you might find yourself replacing: LEAD IS TOXIC. EXPOSURE TO LEAD DUST CAN CAUSE SERIOUS ILLNESS, SUCH AS BRAIN DAMAGE, ESPECIALLY IN CHILDREN. The label also noted, in smaller print: "This product contains chemicals known to the state of California to cause cancer, birth defects and other reproductive harm." It also specified that the paint contained "VOC <50 g/l." "VOC," I guessed, meant "volatile organic compounds," the stuff that people with multiple chemical sensitivity find particularly onerous. These compounds are released into the air, or "volatilized," as paint is spread. But the number? Was that a good number or a bad number? And how was I supposed to know? (When I looked into it later, I found that a VOC level of below 50 grams per liter was actually pretty good. Nonetheless, the EPA notes, during and immediately after the application of paints and lacquers and thinners, levels of VOCs can be *a thousand times* stronger than background levels. Which is not so good.)

So the hardware section, as Albert Donnay could have told me, was plainly a major pipeline into our collective barrel of toxic chemicals. But at least most of the products made some effort to display their ingredients and to offer warnings, however unsettling and vague. The same could not be said of the next section Katherine and I visited, where we found many of the same ingredients we had seen in hardware, but without the warning on the labels. And these products were not meant to be applied with rubber gloves. They were meant to be slathered directly on your skin.

COSMETICS AND PERSONAL CARE

Walking down a cosmetics aisle, if you pay close enough attention, you can actually notice the smells changing with virtually every step you take. As Katherine and I moved from the fragrance section to the powders section to the lotions section, I could sense the changes with my eyes closed; I am not speaking metaphorically.

Not that your eyes are particularly helpful here. As I've mentioned, most labels are not exactly brimming with useful information. It was Katherine who noticed that the children's bubble baths were marked "Keep Out of the Reach of Children." This was a few steps down from a nail polish, boasting to be both "Hi Res" and "Hi Def," that was packaged in a glass container with an ingredient list so microscopic that neither Katherine nor I could read a single word of it—except for "Caution: flammable."

When I showed Katherine lotions called Velvet Tuberose and Decadent Amber, we agreed that people selling these things must have been thinking about sex toys or the names of their favorite strippers. Over and over, her eyebrows rose when she compared the marketing strategies of the skin care stuff targeted at women with that targeted at men. For women, it was all Satin Luxury, Glowing Touch, and Caress. For men: Snake Peel, Instinct, Dark Temptation, Phoenix, High Endurance. The ingredient lists were equally opaque; despite the lengthy enumeration of contents, it was impossible to tell what any of this stuff was made of. But in an ironic twist on a safety warning, Axe bath products marketed to men featured a silhouette on the back label of a man showering with not one but two women. The label cautioned, "The Axe Effect may result in, but is not limited to, unrelenting female attention and/or late nights."

Katherine chuckled as she read it. (Interestingly, she noted, the Axe hair products featured a label showing a man with only one woman. It's not clear why this would be. Perhaps monogamous men are more likely to wash their hair.)

At one point, Katherine pulled down a body lotion billed as containing "shea and cocoa butter"; the third ingredient, far ahead of shea and cocoa butter, was petroleum. She also found petroleum in the deodorants. She examined an "exfoliating body wash." The stuff claimed to be made with "microbeads," tiny bits of plastic that, along with a vigorous scrubbing, help slough off dead skin. When she turned the bottle over, she found that the wash's second ingredient was petroleum. Was that what the microbeads were made of? Or were they made of the acrylonitrile/methacrylonitrile/methyl methacrylate copolymer? It was hard to tell. Either way—and regardless of the effect these bits of plastic might have on warm, damp skin—all I could think of were all those fish out there in the Great Pacific Garbage Patch, swimming around in the countless trillion microbeads that end up flowing down the shower drain. And where these microbeads would end up, once those same fish were caught in someone's fishing net. What goes around comes around.

As for the lotions that make the hair fall off your legs so you don't have to shave, Katherine said she didn't even want to know.

All she could tell from the label was that they were toxic enough to merit a warning about getting the stuff too near your eyes, your nipples, or your genitals. "For a store that's trying to sell to people who are trying to save money," Katherine said, "there's so much here that people just don't need. There's another way to save money: don't buy anything."

The cosmetics section was remarkable mostly for the radical shift in its marketing slogans. There were no "Unbeatable" price signs here. What there were instead were dozens upon dozens of photographs of beautiful, rich, toothy, bronzed celebrities. There were countless claims about products being "natural" (or, in some cases, "naturale"). Others boasted of their products' "science," its powers of "therapy" or "healing." This was as true for the colognes as it was for the shampoos and hair colorings and nail polishes. If you believed the signs, this would be the healthiest section in the whole store.

Which, of course, is just the way the cosmetics industry wants you to see its products, and its marketing is plainly working. The average American applies some form of personal care product (toothpaste, shampoo, moisturizer, lipstick, nail polish) a dozen or more times a day; many of these products contain carcinogens like coal tar and 1,4-dioxane; neurotoxins like lead and mercury; and hormone disruptors like parabens. Fully one-third of women over eighteen in North America, Japan, and Europe (and 10 percent of men over forty) use some form of hair dye, which can contain formaldehyde. All told, the personal care industry sells some 6 billion products a year, for roughly $50 billion a year in revenue.

Do you know what's in any of it? Do you have any idea what's in your shampoo? How about your moisturizer? Your baby's moisturizer? Another way of thinking about this: What do you *think* is in this stuff, and where are you getting your information?

For decades, the health effects of personal care products were of no concern. Newspaper and magazine advertisements in the 1950s urged women to buy Satura, a skin cream boasting that it contained hormones to "plump up skin" and "smooth tiny lines." In the 1980s, ads in *Essence* magazine urged African-American women to buy Le Kair hair treatment with "Hormones Plus Vitamins," and Hask Perm-Aid with Placenta.

But in recent years, researchers have been finding striking evidence that chemicals contained in all sorts of personal care products are altering hormones in both animals and humans. More confounding is the veil that surrounds the ingredients in these products. If the hardware department is where you find the most explicit warnings about toxic chemicals, there is probably no section in a big box store with less information than cosmetics, even though you can find some of the same chemicals at work. Nail hardeners can contain formaldehyde, used as a resin and preservative, and toluene, a clear, colorless liquid that acts as a solvent and helps suspend colored pigments in the polish. Like phthalates, which can be found in everything from soaps to baby lotions, toluene is a possible reproductive and developmental toxicant. And the growing use in cosmetics of so-called nanoparticles— chemicals that are a thousand times smaller than a human cell—is raising a new set of eyebrows. Researchers at the University of Delaware recently found that synthetic nanoparticles are readily taken up by plants (in this case, by pumpkins), can accumulate in their tissues, and can thus readily enter the food chain. Of particular concern seem to be zinc oxide and titanium oxide, used to make products like sunscreen transparent.

A couple of years ago, after leaving a classroom discussion of toxic chemicals in consumer products, a female student of mine went home and wrote a journal entry about looking at her cosmetics shelf with new eyes. Of particular interest was a bottle of nail polish that boasted that it was a "formaldehyde and toluene free formula." She had just learned that the European Union had recently banned some 1,100 toxic chemicals from personal care products and that in the United States, the Environmental Working Group had launched Skin Deep, the world's largest database of chemicals in these products. The database, she discovered, had found that one-third of personal care products contain at least one ingredient linked to cancer; 45 percent contained an ingredient that can be harmful to reproduction or child development; 60 percent contain chemicals that can mimic estrogen or disrupt other hormones; and 89 percent of the ingredients have never been assessed for safety by the Cosmetic Ingredient Review panel.

The young woman found some startling information in the Skin Deep database. Her shampoo, conditioner, deodorant, mascara, eyeliner, and toothpaste were all listed as having "moderate" hazard levels. Her conditioner, for example, contained ingredients that had been linked with cancer and immunotoxicity and were known to linger in body tissue for "years or even decades after exposure."

"Before then, I had never once thought that doing something as seemingly innocent as getting my hands ready for spring could be introducing carcinogens into my body," she wrote about her toluene-free nail polish. "Had it not been for the label stating that it was absent, I would never have realized that it is, in fact, present in other nail polishes. The fact that I know nothing about toluene's effects on the human body does nothing to calm my nerves, and looking at the list of all the ingredients in the polish, none of which I recognized, only made me more anxious."

Research is building that breast growth is starting as much as two or three years earlier than in the past, and that menarche begins as much as a year earlier. There has been a big jump in the number of girls developing secondary sex characteristics by the age of eight, and environmental estrogens may be causing this change. Other studies have shown troubling signs of early-onset puberty in African-American girls, possibly from the use of hair products that contain estrogen or placenta, products that are left unregulated by the FDA. (African-American women are less likely than white women to get breast cancer, but at any age they are more likely to die from it.) In May 2009, CBS News reported on a girl in La Mirada, California, who underwent a mastectomy as part of her fight against breast cancer. She was ten years old. "I feel like she's been robbed of her childhood," the girl's mother said. "It's beyond shocking. You know. She's ten. She has breast cancer. It's unheard of."

And these troubles can arise even at levels of exposure well below accepted norms. In 2000, an independent panel convened by the National Institute of Environmental Health Sciences and the National Toxicology Program found evidence that some endocrine disruptors can affect animal body functions "well below the 'no effect' levels determined by normal testing."

All of this, Theo Colburn and her colleagues write, suggests that over the last fifty years, "synthetic chemicals have become so pervasive in the environment and in our bodies that it is no longer possible to define a normal, unaltered human physiology. There is no clean, uncontaminated place, nor any human being who hasn't acquired a considerable load of persistent hormone-disrupting chemicals. In this experiment, we are all guinea pigs."

This notion of a hormone-disrupting chemicals being inescapable felt truer in the big box store than anywhere I had ever been. They were not just in the products themselves; they were literally in the air we were breathing. Throughout our visit, Katherine and I had not just been astonished by what we'd found on the labels; our noses and our throats had been assaulted by the stuff. Especially that "lavender" smell. We needed to find out where it was coming from.

At first, we had been sure it was coming from the shampoo aisle, but that was more a symphony of "mixed-berry," "tropical fruit," and "coconut." Then we figured it was coming from the home cleaning aisle, but that proved more a source of lemon mixed with chlorine. Finally—at this point I felt like a bloodhound, my nose in the air—we discovered the Lavender Epicenter, right near the intersection of three departments: pets, home, and kitchen. It came in waves I could practically see. Lavender body washes. Lavender foot washes. Lavender potpourri, dried (to be placed in a bowl in a bathroom). Lavender potpourri, liquid (to be placed in a pot on the stove and simmered). Lavender laundry soap. And, most overwhelming of all, lavender-scented candles.

It was here that I almost vomited.

On a shelf of deodorizers. I found a fist-sized air freshener that comes equipped with a motion sensor: you attach the thing to a wall, and it squirts fragrance into the air whenever someone walks into the room. I was marveling at the product's list of warnings (call poison control if ingested; beware of electric shock) when Katherine snuck up behind me and asked me to take a whiff. She held up a glassed-in candle and removed the lid under my nose. What came out was a blast of "peach cobbler" so overpowering that I gagged and almost lost it.

Good God, I said. Get that away from me. I thought of people with multiple chemical sensitivity and how this place, for them, would represent a rather deep circle of hell.

Katherine laughed. She showed me another product, this one an aerosol can of "home-made apple pie." No need to fill your house with "heaping barrels of brightly colored spices," the label said. "Just give it a spray or two!"

As I tried to collect myself, Katherine pondered the meaning of these smells. Selling a product that uses a combination of chemicals to make people imagine a "peach cobbler" or a "home-made apple pie" was selling nostalgia to people too busy or too overwhelmed to get the smell the old-fashioned way: by baking a peach cobbler or an apple pie.

"It's pretty sad that you have to spray the smells that you crave," she said. "What do these smells *represent*? That Grandma is home taking care of you. These things are marketed to overworked people. It's all so nostalgic—for cleaning, and cooking, and comfort and security. Why bake an apple pie when you can spray an apple pie?"

Some of these products, especially the lavender aerosol sprays, were marketed specifically to "deodorize" the home. This is also true of the various sprays on display in the pets section, where the shelves are also full of deodorizers and cleanup products. (I'd like to tell you what was in these products, but none listed a full set of ingredients. The best I can say is that some of them featured a "pleasant cut-grass scent.") Not to state the obvious, but "deodorizing" is one thing these products do not do. What they do is replace—or overlay—one set of smells with another, and the new smells, it turns out, come freighted with petrochemicals and, especially, phthalates.

There they were again: phthalates, the family of chemical compounds added to everything from shower curtains to baby pacifiers to crinkly disposable drinking-water bottles. But phthalates are also added to cosmetics, soaps, and shampoos—almost anything containing "fragrance." In short, phthalates are everywhere. The Natural Resources Defense Council recently examined fourteen air fresheners and found phthalates in twelve of them, including in products advertised as "all natural" or "unscented." More confounding, not a single one had phthalates on its list of ingredients.

Nearly 5 million metric tons of phthalates are consumed by industry every year. For industry, this has been a boon. "The fact of life is that phthalates are a remarkably useful product that . . . allow people without a lot of money to have a first-world lifestyle," Marian Stanley, manager of the Phthalate Esters Panel for the American Chemistry Council, has said. "The risk is a theoretical risk. If you had the smallest baby with the most exposure for the longest time, you theoretically have a risk. Practically, do you have a risk? Nobody's seen it yet."

This doesn't mean that scientists aren't looking. Researchers at the University of Washington and the Centers for Disease Control tested 163 wet diapers from infants born between 2000 and 2005. Then they asked the mothers if they used powders, creams, baby wipes, shampoos, or lotions. Then they asked how many hours a day their babies played with soft plastic toys, teethers, or pacifiers. All 163 diapers contained urine contaminated with at least one phthalate, and more than 80 percent of them were contaminated with at least seven. More than half the mothers said they had used baby shampoo, and more than 30 percent said they had used baby lotion or diaper cream. Infant exposure to phthalates was "widespread," the study reported, and "strongest in infants who were younger than eight months." Especially for people who thought children got exposed to phthalates only by sucking on plastic toys, this was troubling news: infants under eight months were apparently exposed to phthalates through their skin. This was particularly worrisome given the size of baby bodies: a palm full of body lotion is a bigger dose for an infant that it is for an adult. And since infants have not fully developed their metabolic systems, they are not as equipped to shed toxins as adults.

The researchers expressed their frustration that figuring out which products contain phthalates is difficult in part because manufacturers do not specify phthalate contents in the ingredient list. "Parents may not be able to make informed choices until manufacturers are required to list phthalate contents of products," the study concluded. "If parents want to decrease exposures, then we recommend limiting the amount of infant care products used and not to apply lotions or powders unless indicated for a medical reason."

You can see where this is going. Phthalates are used in an unimaginable array of consumer products; people are exposed to phthalates from the moment they apply perfume in the morning to when they take the plastic wrap from the refrigerated leftovers to when they slide into their cars or spray pesticides on their lawns. Phthalates migrate; they end up in our bodies. As with flame retardants, the trouble with phthalates is that they do not bind chemically to the products to which they are added; instead, they leach out into the air as dust, or into liquid, if they are used in things like plastic bottles. Once ambient, phthalates are quick to find their way into our skin, our lungs, and our food. With all these pipes coming in, it's no wonder our barrels fill up fast.

In state after state, phthalates show up in every person who gets tested for them. In the Maine body burden study, phthalates were found in all thirteen participants; three phthalates, including those used to make car interiors, shower curtains, and personal care products, were found in more than 75 percent of all Americans who had been tested for them. In a "Pollution in People" study done in Washington State, ten out of ten participants were contaminated with phthalates. In Oregon, the numbers were the same: ten out of ten people had three phthalates in their bodies. Seven people had six.

In September 2000, CDC researchers went looking for seven types of phthalates—those used in detergents, cosmetics, lubricating oils, and solvents—in the bodies of 289 people. They found at least four in 75 percent of the people tested. But a more revealing statistic emerged when the CDC broke the data down by age and gender. Perhaps because of their use of "perfumes, nail polishes, and hair sprays," women between the ages of twenty and forty had exposure levels five times greater than the average person. "From a public health perspective," the report said, "these data provide evidence that phthalate exposure is both higher and more common than previously suspected."

A study of 85 people published in June 2005 linked fetal exposure to phthalates to structural differences in the genitalia of male babies. Researchers measured phthalate levels in pregnant women and later examined their infant and toddler sons. For pregnant women who had the highest phthalate exposure, their baby sons, on average, had smaller genitalia and were more likely to have incompletely descended testicles. Most striking was that these boys had a shorter perineum, the space between the genitalia and the anus, which scientists call the AGD, for anogenital distance. In rodents, a shortened perineum in males is closely correlated with phthalate exposure. A shortened AGD is also one of the most sensitive markers of demasculinization in animal studies.

Strangely enough, federal guidelines require that phthalates be listed on ingredient labels for things like nail polish, but not when they make up part of the "fragrance" or the "secret formula" for many other products, such as perfumes and hair mousse. And as so often is the case, federal regulation of personal care products seems hamstrung at best. The journalist Nena Baker, using the Freedom of Information Act, discovered that the Food and Drug Administration's Office of Cosmetics and Colors employs only thirty people and has an annual budget of $3.4 million. In Baker's hometown of Portland, Oregon, she writes, the city's office that controls traffic signals has forty employees and a budget of $22 million.

So, once again, we are left with companies policing themselves. If you turn to the website cosmeticsinfo.org, whose name makes it sound like a neutral party, you will have to work to find out that it is in fact sponsored by the cosmetics industry. Much of what you will read there trumpets the strong regulatory relationship the industry has with the FDA. "Right now in laboratories across the country and around the world," the website reports, "experts in chemistry, biochemistry, microbiology, molecular modeling, engineering, formulation and toxicology are at work stringently testing and assessing ingredients and formulas for safety, quality, stability, purity, skin efficacy and performance."

Such rhetoric does not always seem reassuring, either for individual consumers or for the 400,000 people who work in nail salons, where the fumes—not only phthalates but toluene, formaldehyde, acetone, and other volatile compounds—are in the air all day, every day. Nail salons have grown 374 percent in the last decade; in 2005, there were some 57,000 nail salons in the United States, employing just under 400,000 workers—95 percent of whom were women, and 57 percent of whom were "women of color," including many Vietnamese. The customers at these salons are also mostly women—about 95 percent. So what you have, in both employees and customers, is an enormous group of women of childbearing age, all exposed to toxic chemicals thought to damage the reproductive systems of developing children. OSHA and states are supposed to enforce compliance with work safety regulations, but in California, for example, there are only 15 inspectors for 9,000 salons.

My near explosion in the air freshener aisle—and the headache that was not going away—made it pretty clear that I needed a break. I located a bathroom, found a water fountain, and took a long, deep drink. The bathroom itself was pretty nice, I have to admit. It had no exterior doors, which I have always taken to be a good public health measure—no doors means no door handles, which means no spreading of germs. It had hands-free faucets and soap dispensers, another good sign. But especially given the context of my journey, the place was—how else can I say this?—incredibly smelly. Or perhaps I should say it was incredibly "deodorized." Either way, the bathroom was overwhelmingly saturated with the smells of detergents, floor cleaners, and—I'm here to tell you—"lavender."

HOME

After I pulled myself together, Katherine and I decided to take a look at the children's pajamas. Well-meaning relatives had been giving our kids synthetic pajamas as gifts since the day they were born, and we had never once stopped to think what went into making them. Why would we? What could possibly be worrisome about a pair of foot pajamas with polar bears on them? Sure, somewhere in the back of my mind I knew the things were "flame resistant," but did I ever wonder what that meant, or if these chemicals could somehow make it into our bodies' tissues? I can't say that I did. It's not like our kids were eating their pajamas. Katherine, a bit more suspicious than I, usually bought cotton pajamas rather than synthetic ones. She knew she was buying the right ones because the pajama labels would say "not intended for sleepwear." I was never that clever.

In a big box store, the kids' pajamas are arranged according to their tie-ins to movies and television shows. In the girls' section, you can choose from Dora the Explorer, a variety of Disney princesses, and Winnie the Pooh. Boys are offered an even wider selection: Thomas the Tank Engine, Spider-Man, SpongeBob, Buzz Lightyear, and any number of sporty-themed pajamas (though none with any professional team affiliation, since these are presumably protected by licensing agreements). So what does a parent see? A familiar face (Dora, Cinderella, SpongeBob); a cheap price; and a tag, typically in block letters, reading: "Flame Resistant Sleepwear: Sleepwear should be flame resistant or snug fitting to meet U.S. Consumer Product Safety Commission sleepwear requirements." The federal Consumer Product Safety Commission, it turns out, requires that fabrics used to make children's pajamas "must self-extinguish after exposure to a small open flame." Sounds pretty good, right?

You'll find similar tags in the home furnishings aisle, attached to the mattresses and cushions of the swivel chairs. Read the fine print of the pamphlet that comes with a new computer or a new television, and you'll likely discover the same. Although you sometimes may not. Even when products are made with flame retardants, it's not always possible to tell. When my wife and I replaced an old mattress recently, I yanked the label off the old box spring (the one warning that it should not be removed "under penalty of law" by anyone except the consumer). The tag was stamped with almost illegibly small print; when I looked for an ingredient list, all I could find was that it was made of "100% SYN FIB PAD/UNKN KIND." This was not exactly helpful information.

So what's the story with all these flame retardants? Are they beneficial?

When I was a kid, my youngest brother and I slept in a bunk bed on the second floor of our house. One night, asleep on the top bunk, I awoke to a noxious smell coming from below. I leaned over and saw that my brother had fallen asleep with a reading light still burning hot under his blanket. Acrid black smoke rose from the cheap foam mattress. I leapt out of bed, jostled my brother awake, and—in a move that remains part of family lore—heaved the entire mattress out the second-story window. The next morning, my parents—who had slept through the whole ordeal—looked outside and saw the mattress, with a nasty black hole in the middle, out on the lawn. As designed, it had smoked but had not burst into flames.

Did the flame retardants save our lives? Who can say? According to the National Academy of Sciences, each year about 90 people, most of them children, die in fires in upholstered furniture that were started with matches, cigarette lighters, or candles. Another 440 people are injured, and property losses amount to $50 million annually.

Some say that flame retardant chemicals save lives. Others say that foam cushions are among the most toxic objects in a home. It comes down to costs and benefits.

People have been inventing ways to prevent the spread of fire since, well, since people first discovered fire. Over the last couple of centuries, fire prevention has fallen in large part to the chemical bromine. Since its discovery in a salt marsh in France in 1826, bromine—a reddish liquid with three times the density of water—has been used to interfere with the oxidation reaction that causes a material to burn. A couch or a pillow made with untreated foam will burn. Treated with brominated flame retardants, it will only smolder. Its name reflects the way bromine smells: *bromos* is Greek for "stench." For more than thirty years, companies have combined bromine with oxygen, hydrogen, and carbon to make flame retardant compounds known as polybrominated diphenyl ethers, or PBDEs. The most common of these are known by their nicknames: penta, octa, and deca. Though they were banned in Europe in 2004 (and discontinued in the United States soon afterward), penta and octa were used for decades—penta to make blown polyurethane foam for furniture cushions, octa to make injection-molded plastic housings for computers, televisions, telephones, and automobile parts—and products containing these compounds are still very much among us. Deca, which is still being manufactured, is also used for high-impact polystyrene resins for electronic equipment, as well as in upholstery fabrics and plastic furniture and toys.

These additives are not incidental: flame retardants can make up as much as 30 percent of overall material weight, and they have become the second-largest class of additives used by the plastics industry. Penta can account for as much as 30 percent of a cushion, for example. As much as 15 percent of the plastic casing around your television is probably made of deca, as are as much as 27 percent of your upholstery fabrics.

Here's the problem: the chemical structure of these compounds closely resembles the structure of far more notorious compounds like dioxins and furans—the acutely toxic chemicals created during the incineration of plastics—and of PCBs, the fire-resistant fluids found in electrical transformers that were considered so toxic they were banned thirty years ago. They're also structurally similar to thyroxine, the growth-regulating hormone created by the thyroid gland; scientists believe that once they begin circulating in the body, flame retardants may block the movement of thyroid hormones. This, in turn, can affect both physical growth and energy levels and disrupt prenatal development. Thyroid hormones are responsible for regulating many essential metabolic functions and are extremely important in promoting normal brain development in infants. The disruption of thyroid-hormone regulation has frequently been associated with permanent behavior problems and brain damage. Rodents exposed to the chemicals showed an increased risk for liver and pancreatic tumors, and for thyroid cancer.

Flame retardants are one of the most common of a group of chemicals known as persistent organic pollutants. POPs are troublesome not only because of their toxicity but also because they are hardy and because they can travel. Since the compounds do not chemically bind to plastics, they are prone to "migrating," or leaching out as dust. Wipe your fingers along the baseboard of your house, and look at your fingers; you'll find flame retardants. Vaccum your carpets, and you'll find them in the dust. You'll also find them in the lint in your dryer. Researchers have found increased levels of flame retardants in the air of office buildings, especially during business hours. Once thrown away, cushions and computers and television sets will, eventually, begin to break down. But their chemical components, apparently, do not. After being crushed in a landfill and exposed to sun and rain, flame retardants, like everything else, turn to dust. Not a little dust, a lot of dust: each year, 1.75 million tons of carpeting are thrown away. From there, it's into the air and the soil and the water. And from there, it's into our lungs, into the food chain, and into us.

So there's a cost.

As Katherine and I stood pawing through the racks of children's pajamas, I had to wonder. Flame retardant compounds are built into products. So why don't they stay there? Do I need to worry that my kids are wrapped in pajamas treated with these chemicals, and are laying their heads on mattresses made with them, and are watching commercials for these very same products on televisions made with them? Suddenly, all those familiar cartoon faces staring out at me from the pajama racks seemed slightly less inviting.

Toxic flame retardants were first identified in living organisms in 1981, when they were found in fish samples from western Sweden. Since then, they have been found all over the world: in birds, fish, shellfish, amphibians, marine mammals, sewage sludge, sediments, air samples, meat, dairy products, and even vegetables. They have been found in North America and Europe and Japan. They have been found in the Arctic. A Canadian report issued in 2002 found that contamination had become widespread even in wilderness areas; levels were doubling every two to five years in lake trout, herring gull eggs, and beluga whales in the St. Lawrence River estuary.

The quantity of flame retardants detected in people and wildlife appears to have doubled in North America every four to five years for three decades, a pace unmatched by any other contaminant. The chemicals turn up in human blood, fatty tissue, and umbilical cord blood in every region where scientists have conducted studies. In many areas, concentrations in humans have been increasing exponentially.

But it wasn't until flame retardants started showing up in breast milk that people began to pay attention.

Flame retardants are "fat-soluble," which means the body tends to store them in fat tissue. They are thus prone to stick around in a person's body longer than something that is, say, "water-soluble" and that the body readily excretes. So when researchers go searching for flame retardants, they tend to look for them in places like the blubber of seals or the breast milk of women. In 2000, Swedish scientists reported that flame retardants had become widespread in women's bodies. The study surveyed nearly thirty years of research into the presence of long-banned industrial chemicals like PCBs and pesticides like DDT and dieldrin in the breast milk of healthy, nonsmoking women living in Stockholm. Some of what they found was encouraging: over time, it seemed, the presence of several of the compounds had decreased after they were banned in the 1970s. An infant's exposure to banned organochlorine compounds in 1997 was one-tenth what it had been in 1972, the report found.

But when it came to flame retardants, the findings were far more ominous.

The levels of flame retardants in people's bodies had doubled every five years, and were now roughly fifty times greater than in the early 1970s.

As public concern grew, the Swedish government reacted quickly, moving to prohibit the use of PBDEs in late 1999. Simultaneously, large Swedish companies like IKEA and Volvo moved to phase them out as well.

The effects? A follow-up study found that from 1997 to 2000, PBDE levels in Swedish breast milk had dropped 30 percent.

In just three years.

So how flame retardant are American breasts? It turns out that American women have breast milk that is considerably more contaminated than European women, yet regulators have done considerably less to address it. In 2004, a team of researchers led by Arnold Schecter of the University of Texas Science Center studied breast milk taken from forty-seven nursing women in Dallas and Austin, Texas. They discovered the highest levels of flame retardants ever reported—ten to one hundred times higher than levels found in Europe. They found elevated levels in women with no known occupational exposure. The women were white, African-American, and Hispanic, and the research team found no apparent difference in flame retardant levels between those groups.

And if you think those women in Texas were contaminated, consider this: one year later, researchers examined tissue taken from fifty-two people who had undergone liposuction with a New

York plastic surgeon. Since flame retardants tend to accumulate in fat tissue, it was not surprising that some would show up. But how much showed up was stunning. Most researchers find flame retardants in human blood and milk at levels between 4 and 400 parts per billion. A thirty-two-year-old man had levels of nearly 10,000 parts per billion. A twenty-three-year-old woman registered over 4,000 parts per billion. Linda Birnbaum, then the director of the EPA's National Health and Environmental Effects Research Laboratory, called the findings "a big shock." And skinny people should take heed: since toxins can be diluted by greater quantities of fat, people with less fat may in fact have higher concentrations, the study said.

As you can see, once you start thinking about flame retardants, it's a pretty quick jump from children's pajamas and sofa cushions to breast tissue. But there's another link in this chain that we need to think about, and that has to do with food. Studies in Japan and Europe have shown a link between flame retardant levels in women's breast milk and the amount of fish in their diet. Farmed salmon turn out to have "significantly higher" levels of flame retardants than wild fish, likely because they are fed ground-up fish that are themselves contaminated. And since farmed salmon has become so popular, the risk of broad exposure to flame retardants continues to increase. In the last two decades, global consumption of salmon has risen from 27,000 tons to more than 1 million tons annually.

All of this news compelled Arnold Schecter and his team in Texas to push their shopping carts through three major supermarkets in Dallas and take a good, long look at thirty different kinds of food—mostly meat, fish, and dairy products. They found levels of flame retardants that were as much as twenty times higher than levels reported for food from Spain or Japan. They even found deca in soy infant formula; to the best of his knowledge, Schecter reported, this was the first time that a persistent synthetic toxin had ever been found in a vegetable product. How does soy formula get contaminated with flame retardants? Was the crop contaminated or did the formula get tainted when it was processed or by its packaging? If you're a mother, does it matter?

And even if you feel confident that a single exposure to toxins in food will not cause you any problems, what do you say, walking through a big box store, when products containing these chemicals fill your shopping cart? How about a lifetime of shopping carts? And what happens when these chemicals, once inside you, mix with the chemicals you get exposed to from, say, your paint? Or your cosmetics? Or, more directly, when they simply release their toxins straight into your children's mouths?

TOYS

Big box stores are frequently criticized for serving as the end point of a Chinese assembly line, and no section reinforces this idea as vividly as the toy department. But as Katherine and I made our way into the toy aisles, it occurred to me that this is not, in fact, the end of the story. All of this stuff—in the toy, home, and hardware departments, everywhere—will one day pass through our homes, end up in landfill or in the air, and begin breaking down. At least some of the ingredients will turn into airborne dust, or particles filtering up through the food chain, or molecules of volatilized aerosols, and will end up in our bodies. It's just a matter of time. But somehow, the link between plastic toys and toxic chemicals seems particularly insidious—and, in recent years, particularly common.

We've all spent time in these aisles, flanked by plastic toys from floor to ceiling. In the store Katherine and I visited, virtually every duck, dinosaur, princess, and superhero was made out of plastic and had been manufactured in China. I'm not talking about a handful of toys in a small corner of the store. I'm talking about *acres* of toys. As in the pajama section, a sizable majority of the products were movie or television tie-ins: *Star Wars*, *Batman*, Ultimate Fighting action figures, Thomas the Tank Engine trains. Others were slightly more original. During our visit, Katherine looked at a product called the Pink Princess Stepstool Potty, which claimed to play four "royal" tunes as a "reward." She looked bemused, but then her eyes darkened. Though we had never bought a musical toilet, we had offered our daughter rewards for learning to use the toilet. We had taken her to a store not unlike the one we were now in, and encouraged her to pick out a small toy. She had chosen a play umbrella. Some weeks later, rewarded again, she had picked out a pair of play gardening gloves. Six months later, both toys were recalled because they had been coated with lead paint. So much for enlightened parenting.

Just two years before our big box safari, the United States experienced what will, in some circles, be remembered as the Year of the Toxic Recall. In the spring of 2007, Chinese-made pet food was recalled because it contained melamine, a toxic synthetic used in cleaning products, plastics, fertilizer, and paint; it killed some four thousand American dogs and cats. In China, melamine found in baby formula and in chicken eggs killed four babies and sickened tens of thousands of others.

In May, almost a million tubes of Cooldent toothpaste imported from China were recalled because they contained diethylene glycol, a solvent, typically found in antifreeze, that can lead to kidney damage or liver disease. The toothpaste had been distributed to jails as well as to expensive hotels. No deaths were reported from the tainted toothpaste, but Panama claimed that hundreds of people died, and hundreds more were poisoned, after using cold medicine imported from China that contained the same chemical, diethylene glycol.

But the Year of the Toxic Recall will be remembered most for an astonishing number of contaminated toys. In June, a toy eyeball made in China was recalled after it was found to be filled with kerosene. Around the same time, RC2, the company that makes Thomas and Friends wooden railway toys, recalled 1.5 million toys because of lead contamination—a small number, really, when you consider that in the previous three years some 20 million pieces of children's jewelry had been recalled because of lead. In August, Toys "R" Us offered customers refunds for about 1 million baby bibs, after tests showed that some, made in China, had also been contaminated with lead. A month after that, Target recalled 350,000 gardening toys because of lead, and Target and RC2 announced the recall of an additional 550,000 train toys, also because of lead contamination. RC2's chief executive apologized for the "burden that recalls create for parents."

A burden indeed. Most Americans thought lead had disappeared as a consumer threat thirty years ago, when it was phased out of gasoline and banished from things like paint. But in the Year of the Toxic Recall, this was just the beginning. The day after the Target recall, a group called the Center for Health, Environment and Justice tested fifty plastic toys and discovered high lead levels in eleven of them. Ten were made of PVC plastic, and three others contained "extremely high lead levels": a Go Diego Go backpack, a Superfly monkey, and a pair of Circo Lulu boots. The backpack had lead levels nearly eight times the health standard set by the Consumer Product Safety Commission.

"It's absolutely astonishing to us that lead continues to be found in children's toys, despite the fact that consumer and environmental groups have been warning the government about the issue for more than ten years," a group spokesman said.

So what was going on? In 1973, car fumes were considered the biggest source of lead in the air, dust, and dirt. A year later, under provisions of the Clean Air Act, the EPA set a timetable to phase out lead in gas, and within a few years average lead levels in children plummeted. What was it doing, creeping back into our lives like a shadow from the past? A toy train is made of wood (or plastic) and a bit of rubber and metal, all held together with glue and painted green or blue or red. Pretty simple. But what's in the plastic? What's in the paint? What's in the glue? What happens when the whole thing is in your child's mouth? Do you have a right to know the answers to these questions?

Children put toys in their mouths that are "intended" to go in their mouths, and they put things in their mouths that are not intended to go in their mouths. What we've discovered, in recent years, is that neither category guarantees that a product is made without hazardous chemicals.

The hazards of lead have been described in medical writings for two thousand years, and concerns over lead-contaminated water were acknowledged in England as far back as the late 1700s. Lead is known to be toxic to the reproductive system, the liver and kidneys, and the immune system. Studies throughout the 1970s and 1980s showed that even low-level lead poisoning can cause neurological damage in the developing brain, and that children exposed to lead are at higher risk of developmental deficits. And although the link between lead and cancer in humans is less clear, lead has been shown to cause kidney and brain tumors in animals. Recent studies indicate that lead may increase the risk of cancer by compromising a cell's ability to protect or repair its DNA that has been damaged by other chemical exposures.

Despite scientific warnings about lead that have been voiced for decades—or centuries, depending on how you look at it—lead poisoning continues to be a significant public health concern. In this country, lead has been banned from paint since 1977, but in the late 1980s 3 to 4 million children were still being poisoned each year. Getting rid of lead proved remarkably effective: during the phaseout of leaded gasoline, women of childbearing age experienced dramatic decreases in blood lead levels, and the number of children under six years old with elevated blood lead has declined from 88 percent in the 1970s to 1.6 percent in 2005.

These changes were hard-won, but lead remains a major problem, especially for the urban poor; African-Americans and other nonwhites have higher mortality rates at lower blood lead levels than whites. Ellen Silbergeld, a professor of environmental health sciences at Johns Hopkins who has been awarded a MacArthur "genius grant" for her work on lead poisoning, has been particularly critical of the lead industry. Big Lead, she writes, has been as sophisticated as Big Tobacco in deflecting blame for the health impact of its products—in this case, by blaming lead poisoning on "decaying cities," and thus on the people who live in them. Lead poisoning is called a "social problem," in which "mothers and children [are blamed] for causing lead poisoning: mothers for their neglectful or ignorant childrearing practices, and children for being stupid a priori and thus prone to eat lead paint," Silbergeld writes. Some eight decades ago, Felix Wormser, the secretary of the Lead Industries Association, once referred to the poor, black, and lead-poisoned children of inner-city Baltimore as "little rodents."

Dr. Silbergeld, whose work figured prominently in the push to ban lead from gasoline, will have none of this. "These defenses," she has written, "avoid the obvious: only lead—not inadequate mothers, stupid children, or blighted cities—can cause lead poisoning." And new studies back her up: in Ohio, researchers at the University of Cincinnati followed hundreds of poor urban children from childhood to maturity, and found that exposure to lead can lead to measurable losses in the very parts of the brain that control impulses, emotions, and judgment. The practical result? For every 5 microgram increase in a person's blood lead level as an infant, their chance of being arrested for a violent offense later in life jumped 30 percent.

More recently, Dr. Silbergeld has found that lead can cause a number of problems for women in many stages of their adult lives, including pregnancy, breast-feeding, and during menopause. Lead, like calcium, is stored in the bones, and during these stages both can be released into the bloodstream. As a woman's bones begin to thin during menopause, her blood lead level can rise as much as 25 percent, causing a rise in blood pressure, diminished cognitive skills, and declining kidney function. Men can suffer from leaching lead as well; studies have shown that men with higher blood lead levels have lower scores in manual dexterity, decision making, and verbal skills. Other research has begun asking whether early exposure to lead might worsen the cognitive decline associated with aging.

Yet during the Year of the Toxic Recall, it wasn't just mothers in cities who were worried about lead poisoning. It was everyone. If American laws had removed lead from paint and gasoline but we were buying our toys from a country with less regulatory muscle than our own, had we traded one peril for another? For a child, the danger of lead paint is the same, after all, whether the paint is flaking off a windowsill or sucked off a toy train.

Here, at least in part, is what happened. In just the last decade, imports of Chinese consumer products nationwide have surged from $62 billion to $246 billion. Nearly 20 percent of the consumer products for sale in the country today are Chinese-made, compared to 5 percent in 1997. China now makes roughly 80 percent of the toys sold in the United States, and the toy companies that sell them here are largely—does this sound familiar?—self-regulated. A month after the recall, RC2 launched a "Multi-Check Safety System" to reassure consumers that the company's products were safe. This would include greater scrutiny of its manufacturing partners, frequent lab tests of paint, and random plant inspections. "Going forward, the truest measure of our success is and always will be the trust parents place in our products," the company reported on its website. "We are working very hard to preserve and strengthen that trust. And, we hope that you and other parents recognize that our products have been subjected to intense internal scrutiny and testing and are the safer for it."

In the Frequently Asked Questions section on its own website, the company posted this: "In the United States, toys and other articles intended for use by children are required to meet a regulatory standard for lead (16 CFR, Section 1303) established by the U.S. Consumer Product Safety Commission in 1977. The standard requires the concentration of 'total lead' in any children's product not to exceed 600 parts per million (PPM). In addition, the American Society for Testing & Materials (ASTM) has published a voluntary consensus standard (ASTM Standard F963-07) that establishes a 90-PPM maximum for what's called 'soluble lead.' While compliance with this standard is not obligatory, most U.S. manufacturers, including RC2, take steps to ensure that their products comply with this standard as well as with the mandatory standard for total lead."

Do you find this reassuring? Can you even tell what it means? Are you heartened or anxious to learn that products are subjected only to "internal scrutiny" and that federal standards are "not obligatory"? Do you consider the list of benchmarks—the ASTM's standards for PPMs—enlightening or perplexing? It's easy to understand why a manufacturer might feel comfortable with a "voluntary consensus standard" on toxic chemicals. But how about consumers? And what is the ASTM, anyway?

These thoughts had just begun to form in my mind when, during our walk through the toy section, I was startled by manic laughter coming from a shelf just below my knee. When I looked to see what was making such a racket, I noticed the spinning, cackling head of a "Royal Giggles Cinderella." The laughter did what it was apparently supposed to do: Katherine picked the toy up.

"Made in China," she said. "Let's see if it has an ASTM label. Yep, there it is: 'Conforms to ASTM F963.'

"I always look for the ASTM label when I buy toys for our kids, or for other kids' birthdays," she said.

"What does that label mean?" I asked.

"To be honest, I don't really know," Katherine said. "I figured if a toy qualified for the label, it must be safe. Now I'm pretty sure that most of the toys I bought were later recalled."

A few hours later, I looked at the ASTM website and found that it was once known as the American Society for Testing and Materials. Formed a century ago to address faulty tracks in the nation's growing railroad system, the ASTM is now "one of the largest voluntary standards development organizations in the world." The group serves as a "trusted source" for setting production standards for an unimaginable array of industrial materials and consumer products. ASTM F963, to which the cackling Cinderella conforms, "relates to possible hazards that may not be recognized readily by the public and that may be encountered in the normal use for which a toy is intended or after reasonably foreseeable abuse." If I would like to order the full fifty-six-page list of standards for this kind of product, I was informed, I could do so for $60.

What federal oversight there is for toys—and it isn't much—is offered by the Consumer Product Safety Commission, a beleaguered regulatory body that, ever since it was created by Congress in 1972, has been seen as an onerous bureaucratic obstacle by manufacturers and business groups. Until last year, it hadn't adopted any major regulations in nearly two decades. The staff in 2008 was just four hundred workers—half what it had been in the 1980s. Charged with monitoring the unimaginably wide conveyor belt of goods coming into this country every year (in 2007 it was worth $614 billion) is a staff of—wait for it—fifteen people.

Not only that, but in the two years prior to the Year of the Toxic Recall, the Bush administration cut the budget and the number of inspectors overseeing the country's vast network of ports, warehouses, and stores. The commission's shrinking budget was just $62 million in 2007, even though the agency regulates an industry that sells $1.4 trillion worth of toys, tools, and televisions every year. The Food and Drug Administration, with a $2 billion budget, spends nearly twice as much monitoring the safety of animal feed and drugs. All told, the CPSC has just 81 field inspectors, who work out of their homes, compared with a network of field offices with 133 employees in 2002. Among their missions? Oversee tens of billions of dollars' worth of toys and other consumer products sold in the country each year. Here's what that looks like: in Los Angeles–area ports, through which 15 million truck-sized containers move a year, a single agency inspector, working two or three days a week, spot-checks incoming shipments.

Here's what else it looks like: when The New York Times sent a reporter to the agency's product-testing lab, which operates out of a former missile-defense radar station in Gaithersburg, Maryland, he found a single lab worker using a magnifying glass and a mechanical stopwatch to help conduct a fabric flammability experiment—the same equipment she had used for three decades. The toy laboratory, down the hall, was an office so cramped that the only space dedicated to a drop test to see if toys will break into small pieces and cause a choking hazard was the spare space behind the office door. "This is the toy lab for all of America—for all of the United States government!" said Robert L. Hundemer, the one agency employee who routinely tests toys, as he held his arms up in the air. "We do what we can."

In the absence of adequate disclosure by industries or oversight by government regulators, some of the most dramatic recent reporting on toxic products has been done by several of the country's leading newspapers, which have in the toxic chemicals story a chance to do what newspapers have traditionally done best: shine a light on information that is usually kept in the shadows. This is a fact you either find reassuring, if you are an old-fashioned believer in the critical role the press plays in a democracy, or deeply depressing, if you consider the speed with which newspapers are shrinking, laying off reporters, and going out of business altogether.

Wherever newspaper reporters bothered to look, they saw toxic chemicals. And when they turned these products over, they saw China.

In the middle of the Year of the Toxic Recall, Jane Kay set out to look into toxic chemicals for the *San Francisco Chronicle* (a paper that itself was recently losing a million dollars a week in 2008 and is in danger of closing). Kay picked a selection of sixteen common baby toys, had them tested for phthalates and bisphenol A (BPA), and found some startling results. A rubber ducky sold at Walgreens contained a carcinogenic form of phthalate known as DEHP at thirteen times the city's health limit. The face of a Goldberger Fuzzy Fleece baby doll contained one phthalate at twice the health limit. The ring on a Baby Einstein rattle contained BPA, as did a Walgreens-brand baby bottle decorated with colorful fish and the plastic covers on two Random House waterproof books (*Elmo Wants a Bath* and Dr. Seuss's *One Fish, Two Fish, Red Fish, Blue Fish*). The books also contained phthalates.

Around this time—November 2007—the *Chicago Tribune* began running a series of stories on dangerous children's products that would win the paper the year's Pulitzer Prize for investigative reporting. Reporters for the *Tribune* tested some eight hundred children's toys from major city retailers, supermarkets, and discount outlets, and discovered that many of the stores routinely sold toys contaminated by lead paint—some with levels more than ten times the government safety limit. Tests done at the University of Iowa found a fourteen-inch-tall Godzilla with lead levels of 4,500 parts per million, more than seven times the federal and state legal limits. There were twenty-nine recalls of children's products because of lead in October 2007 alone—the highest monthly total since the Consumer Product Safety Commission was founded in the 1970s, the paper reported. The "vast majority" of the toys had been made in China.

The *Tribune* found that even award-winning toys, like colored blocks sold by the Baby Einstein company, had lead levels more than twenty times the Illinois safety limit, which, along with California's, is among the nation's strictest. An inflatable yellow disk showed lead levels twenty-seven times higher than the state limits. A Christmas figurine sold at Walgreens, the nation's largest drugstore chain, had red paint with lead levels eight times the state limit. Yet several toys considered unsafe by state standards were considered safe by federal standards, because of different regulations for lead used in vinyl products.

All this news—again, prompted in large part not by voluntary disclosure on the part of industry but through research done by reporters laboring away in a dying newspaper industry—finally led, in 2008, to some important changes in consumer product safety regulations. The Consumer Product Safety Improvement Act bans lead and a handful of phthalates from children's toys; requires toys to be tested by third-party laboratories before being sold; increases the CPSC's oversight budget; and has created a searchable database of consumer complaints about a product's safety. The database was a particularly contentious point; in the past consumers had been forced to file Freedom of Information Act requests just to get their hands on this information, and companies could block their requests.

During the back-and-forth, industry giants got some concessions. The conglomerate Mattel, Inc.—which in the last few years has recalled more than 14 million toys for, among other things, lead paint—convinced Congress to allow large toy makers to test their own products, rather than submitting them to independent labs. Once again, consumers will be left to trust in corporate self-policing.

The Mattel Loophole, continuing the American tradition of industry self-regulation, remains in place. And the law is full of real-world problems. There are plenty of other phthalates still available for use in children's toys, and there are still plenty of products—plastic shoes, for example—that are not considered toys, and are thus not covered. The law does not apply to jewelry as long as the jewelry is not "intended" for children under twelve. In other words, a plastic ring marketed to thirteen-year-olds is not covered, though it stands an equally strong chance of ending up in the mouth of a baby. And the law doesn't apply to plastic packaging (plastic bags, bubble wrap), which, as any parent knows, can be every bit as intriguing to a child as the product itself. And play cosmetics—commonly made with phthalates—are covered only if they are packaged with toys. If the cosmetics are sold separately, they are subject to completely different regulations, overseen by the FDA. Which, as we have learned, means they are not covered at all.

THE KITCHEN

Of all the sections in the big box store, none got Katherine more worked up than the "food" aisles. I put "food" in quotation marks because every last bit of it—with the exception of some quarts of milk and a few dozen eggs—had been as processed as any piece of plastic in the store. Aisle after aisle was stuffed with soft drinks and snack cakes, cookies and potato chips, Pop-Tarts and doughnuts and ice cream. Even the few "ingredients" for sale—five-pound bags of sugar and bottles of corn syrup—seemed to be marketed at people trying to create processed foods at home.

"It's like a five-year-old went into a food store and waved a magic wand," Katherine said.

The parallels between the food section and the rest of the store were hard to miss, and raised difficult questions. If you are what you eat, this thinking goes—if you are what you put in your body—then aren't you also what you put on your body? Or what, in your home, you use to surround your body? In another parallel with the food industry, a discussion of the products sold in big box stores inevitably brings up questions of class. Like organic foods, nontoxic products tend to be more expensive. Does this make them an indulgence available only to the wealthy? Is it reasonable to expect people to spend more money on nontoxic products when they can get similar products cheaper in a big box store? By extension, is it fair to criticize a business that caters—first, last, and always—to people trying to save money?

But what if the question is asked another way? What are the ethics of limiting a person's consumer choices to products made with unhealthy ingredients? If someone living in a poor city neighborhood can choose only between eating at a fast food restaurant and from a convenience store, their chances of becoming obese and diabetic skyrocket. Is that a fair trade? So what do we say about the impact on people's health from the myriad products for sale in a big box store? Take, for example, nonstick cookware. Of all the brands pushing their products, only one line boasted of its green bona fides. "No PFOA. No lead. No cadmium," the label on one pan claimed. As interesting as the claim was (it was, after all, one of the few products in the entire store that specifically claimed *not* to be made with toxic chemicals), the packaging could not help but make a shopper wonder about what had gone into the other brands on the shelf. If this pan wasn't made with heavy metals or PFOAs (the so-called Teflon chemicals that were killing lovebirds in Maine), what did that say about all the other pans? Were they made *with* them? And if those pans *were* made with heavy metals or PFOAs, and this one was bragging that it wasn't, did that mean that those pans were somehow dangerous? And if you only had a few dollars to spend, which one would you choose?

The toxic chemicals associated with food preparation have received considerable attention in recent years, and I'm not talking about the pesticides sprayed on vegetables. I'm a pretty strict consumer (and grower) of organic produce, so my concern here was not about what gets sprayed on my food. My concern was about how food gets packaged.

Katherine is a fine cook, and we both do everything we can to

eat as locally and organically as we can. In Maryland, thankfully, this is pretty easy. Seven months of the year we get fresh, organic produce from a local organic farm. We have a modest-sized vegetable garden of our own, which produces an array of heirloom tomatoes, collard greens, okra, and peas. But winter is tough. In winter, we turn to canned vegetables.

One of Katherine's favorite dishes involves a homemade tomato sauce (the secret is in the shredded carrots). In cooler months, when our garden is moribund, she typically uses cans of organic, fire-roasted crushed tomatoes from a company called Muir Glen that feature a picture of a tan, smiling, polo-shirted man in a field of organic vegetables. "Inside this can are the richest, sweetest tomatoes you've ever tasted," the label reads. "We're like winemakers and you're holding our pride and joy. All that flavor and goodness came from just the right fertile soil, just the right water, just the right California sunshine. If you love the photograph of the organic tomatoes on the front, we hope you'll agree that they taste just as wonderful as they look."

A big green symbol on the label says the can is recyclable. Another one says that the tomatoes are "USDA Organic," and a third that the contents have been packed in "lead-free enamel." A perfect product in a feel-good container.

Yet something still nagged. One recent winter, Katherine became curious—not about the quality of the tomatoes, which she loved, but about the can itself. She had been hearing stories on the radio about the plasticizer bisphenol A showing up in unexpected places—water bottles, apple juice containers, even the inside of cans used for vegetables. She decided to find out what was lining the cans of her favorite organic tomatoes.

There was nothing on the label about bisphenol A, so Katherine picked up the phone and called Muir Glen. After being patched through a couple of times, she spoke to a woman at the company who confirmed that, yes indeed, cans of Muir Glen organic tomatoes were lined with BPA.

Huh. What do we do now? Given a growing body of evidence that says toxic chemicals leach out of containers into whatever the containers hold, should the tomatoes still be considered organic? Should food containers be required to tell consumers not only what is in the food but what goes into the containers themselves? Are we satisfied with a system that requires consumers to do their own research to find these things out?

And what, exactly, is bisphenol A, and why had we never heard of it before? Even Katherine, who had gone to medical school, had never heard a peep about the stuff. Granted, that was ten years ago, when medical students were told to ask patients if they kept a gun in their homes. No one ever trained them to ask about bisphenol A (or any of the other toxins in their homes, for that matter, save, perhaps, lead paint). In response to customer demand, General Mills, which owns Muir Glen, announced in April 2010 that it would be switching to metal cans made without BPA. The company, however, said it had not found BPA alternatives for many of its other products.

What were we to think about the fact that 95 percent of Americans tested by the CDC had traces of BPA in their bodies, and that the compound has been linked to infertility, genital tract malformations, and increasing cancer rates, especially breast cancer? Or that it was a common ingredient in all kinds of baby products—like baby bottles? And apple juice containers? And the Nalgene water bottles I had been drinking from for twenty-five years?

For years, Katherine and I had taken the kids on road trips every summer to visit friends and family from Virginia to Maine. Being more or less "green," we always brought our own water bottles with us: Nalgene bottles for us, plastic baby bottles for the kids. If a trip lasted more than a day, as they often did, we would leave the bottles in the car. If the trip happened to be during the summer, which they often were, the bottles (and the water inside them) would heat up in the car. If we were in a rush, as we often were, we would forget to refill the bottles, and just pass them around the next day. The water was always warm, and tasted funny, but it was right there.

Like Amy Graham, who, with her husband, had used Nalgene bottles along the Appalachian Trail, I'd always considered these containers the safe way to go. I'd used them for years on backpacking and canoeing trips; they are bigger and more durable than other bottles, and their looped caps make them easy to hang off a backpack or cinch to the thwart of a canoe. Best of all, Nalgene bottles are perfect for carrying hot tea, a luxury in the outdoors. For years, Nalgene bottles have also served as a subtle symbol of undergraduate hippieness; most of my "green" students have them hanging from their backpacks. They're reusable, so they don't end up in the garbage. And they aren't made with phthalates, like all those crinkly disposable bottles.

But a couple of years ago, some mountaineers began asking questions about the plastics that were used in Nalgene bottles, and their questions reached the ears of a group of researchers at the University of Cincinnati. The scientists bought one batch of Nalgene bottles from a local outfitter, then went to the local climbing gym and borrowed a batch of others that had clearly been well used.

What they found was startling. Nalgene bottles were traditionally manufactured with bisphenol A, which makes the polycarbonate bottles stiffer than bottles made with phthalates. This is why they're so durable, and so good for carrying hot tea. The trouble is, the bisphenol A leaks from the plastic into the liquid, and it leaks fifty-five times faster when the bottles are filled with hot liquid.

Bisphenol A has long been known to be an endocrine disruptor, and has been shown to affect reproduction and brain development in animals. Even before the Nalgene experiment, studies had shown that repeated scrubbing or washing released BPA from bottles—a fact that sent chills through parents used to feeding their babies warm milk in plastic baby bottles (with nipples frequently made with phthalates). But the Nalgene experiment was the first to show how dramatically heat speeds up this leaching. The study found that nine-year-old bottles released the same amount of BPA as brand-new bottles. So it's not like your bottles get cleaner with age. They just keep leaching.

Where else can you find BPA? As with phthalates, the answer

is: pretty much everywhere. It's now one of the highest-volume chemicals produced worldwide: global BPA capacity in 2003 was 2,214,000 metric tons, or more than 4.9 billion pounds, and demand has been growing 6 to 10 percent a year. Between 1980 and 2000, industrial production of BPA jumped fivefold.

BPA is used to harden cell phones and laptop computer cases. It lines water pipes. It's even used in dental sealants. A chemical derivative of BPA called bisphenol A diglycidyl ether (BADGE) is used to make epoxy resins that are found widely in the linings of metal food and drink cans. In 1995, researchers in Spain found that the BPA that lines food cans leaches not only into the liquid in the can but into the vegetables themselves. Peas, artichoke hearts, corn, mushrooms, green beans—they all pick up BPA. The data "strongly suggest that some foods preserved in lacquer-coated cans acquire estrogenic activity," the study said.

And the effect of all this plastic in our lives?

"There is a large body of scientific evidence demonstrating the harmful effects of very small amounts of BPA in laboratory and animal studies, but little clinical evidence related to humans," Scott Belcher, one of the Nalgene study's authors, said. "There is very strong suspicion in the scientific community, however, that this chemical has harmful effects on humans."

Like countless researchers before them, Belcher and his colleagues acknowledged that the links between Nalgene bottles and health problems are not necessarily direct. But they warned that the cumulative effect of long-term exposure to BPA and other toxins may be dire.

As with flame retardants and phthalates, one problem with assessing the harm of BPA exposure is that there are so few long-term studies of children—or anyone else—exposed to these chemicals. The federal government's eighteen-year-old guideline for BPA considers it safe for a nine-pound baby to swallow 200 milligrams of BPA per day. Last year, researchers at Tufts University exposed pregnant lab rodents to BPA levels proportionately two thousand times *lower* than that—and found changes in their mammary glands. In humans, such changes are associated with a higher risk of breast cancer. Other research has shown that BPA is connected to the early stages of prostate cancer.

And as with other chemicals, the research into the effects of low-dose exposure to BPA is fairly young. This, of course, allows groups like the American Chemistry Council to question links between the low-level presence of these chemicals and disease— even to the point of speaking directly to women of childbearing age.

"As a mother, I can understand the concern with the confusing and contradictory information about the safety of plastic beverage containers and cans. But as a scientist, I have confidence in what the science says," said Sharon Kneiss, vice president of the products division for the American Chemistry Council. "As the manufacturers of plastic products, we are committed to the safety of our products. We owe it to the public to correct the inaccuracies and mischaracterizations about plastic bottles and the materials used in their manufacture."

As they had done so many times before, newspaper reporters set to work trying to untangle all the competing rhetoric. In March 2007, the *Los Angeles Times* reported that a Bush administration agency responsible for determining the dangers of toxic chemicals to reproductive health had hired a private company with close ties to the chemical industry to determine the safety of BPA. A month after the story surfaced, the company was fired. A few months later, the Milwaukee *Journal Sentinel* reviewed 258 scientific studies of BPA and found that "an overwhelming majority" showed that the compound has been linked to breast cancer, testicular cancer, diabetes, hyperactivity, low sperm counts, and "a host of other reproductive failures" in laboratory animals. Of equal concern, the few studies that found bisphenol A to be benign were paid for or partially written by scientists hired by companies like Shell Chemicals and Dow Chemical. Two others, paid for by General Electric (which made BPA until two years ago), did not even undergo peer review.

In a follow-up report a year later, the *Journal Sentinel* tested ten products marketed for infants or considered "microwave safe" and found bisphenol A leaching from all of them at levels that can cause neurological and developmental damage in laboratory animals. The newspaper found BPA not just in hard plastic bottles but in frozen-food trays, microwaveable soup cans, and plastic baby-food packaging.

In May 2008, Toronto's *Globe and Mail* reported that traces of BPA had been found in every one of fourteen samples of canned goods, with levels especially high in foods often consumed by children, including tomato sauce, chicken noodle soup, and apple juice. "These results provide further evidence that Canadians are marinating in this chemical on a daily basis," Rick Smith, executive director of Environmental Defence, a Toronto advocacy group that has been lobbying Health Canada to ban bisphenol A from food and beverage containers, told the paper. That same spring, Health Canada said it planned to add BPA to its list of toxic chemicals and would soon ban BPA from baby bottles, thus making Canada the first country in the world to take precautionary action against low-level BPA exposure. The decision was part of a larger, nationwide plan to review some 23,000 "legacy chemicals" and concentrate immediately on the health risks associated with 200 problem chemicals found in the environment and in consumer products. Soon after Canada's announcement of a ban, several corporations—including Nalgene, Walmart, Toys "R" Us, Playtex, and CVS pharmacies—said they would stop producing and selling certain products made with BPA.

"In Canada, at least, I can tell you that people are voting with their feet and voting with their pocketbooks, and indeed, for the past several weeks or months now, there's been practically zero demand for infant baby bottles that have BPA in them," Canada's health minister, Tony Clement said. "You know, this is how you get new moms to buy your products, saying, 'Buy here, they don't have any BPA in them.'"

More than a decade ago, the Japanese reduced the amount of bisphenol A in the lining of food cans by 95 percent. Until American companies do the same, Frederick vom Saal, a leading researcher on the effect of synthetic chemicals on human reproduction, has said, "I eat nothing out of cans."

CLEANING PRODUCTS

It was time to move on. Katherine and I left the kitchen section and made our way into what might as well have been called the cleanup department. I could tell we had arrived when the smells shifted, temporarily, from "lavender" to "lemon." Or, more accurately, "lemon swimming pool." We were suddenly in a kind of chlorine wetland.

As I had in the bug-killer department, I felt momentarily overwhelmed by the array of choices. I knew most of this stuff contained chemical compounds designed to kill things—in this case, not insects but microscopic bacteria. I picked up a large white bottle and held it before my eyes. Right on the front of the label was the first warning, in all-capital letters: CAUTION: EYE IRRITANT. MAY BE HARMFUL IF SWALLOWED. SEE CAUTION ON BACK PANEL. I flipped the bottle around and read, KEEP OUT OF REACH OF CHILDREN. CAUTION: DO NOT GET IN EYES. DO NOT TAKE INTERNALLY. FIRST AID—FLUSH EYES WITH WATER 10–15 MINUTES. IF IRRITATION PERSISTS CALL A PHYSICIAN. INTERNALLY—GIVE LARGE AMOUNTS OF WATER OR MILK. CALL A PHYSICIAN.

What was inside this bottle that made it so dangerous? It was hard to tell. Under "Ingredients," it said only "cleaning agents (anionic and/or nonionic surfactants)." That was it. The label then noted that the detergent contained no phosphates. "The surfactants (sudsing/cleaning agents) in this detergent are biodegradable and break down naturally into simpler compounds, helping to eliminate suds and foam from lakes and streams."

So the stuff won't add foam to rivers, but it might hurt my eyes. Do I use it? What happens when it gets in my lungs?

The next bottle was a 96-ounce jug of bleach, which boasted that it "Cleans and Deodorizes, Removes Stains, Eliminates Odors." This one also had a warning on the front of the label: DANGER: CORROSIVE. SEE BACK PANEL FOR ADDITIONAL PRECAUTIONARY STATEMENTS. KEEP OUT OF THE REACH OF CHILDREN. And again, this bottle alerted me to the product's active ingredient: sodium hypochlorite (3 percent), in addition to other ingredients (the rest). What these other ingredients were, it did not say.

What the label did say was that the bleach was toxic. On the back, some two hundred words were dedicated to a list of problems the product might cause, including: "severe skin and eye irritation or chemical burns to broken skin. Causes eye damage. Wear safety glasses and rubber gloves when handling this product. Wash after handling. Use with adequate ventilation."

Farther down, the label advised consumers to call poison control for treatment advice and not to induce vomiting unless advised to by poison control or a doctor. Then it warned that the bleach is a "strong oxidizer." "DO NOT use or mix with other household chemicals, such as toilet bowl cleaners, rust removers, acid or ammonia containing products. To do so will release hazardous gases."

Then, as usual: Keep out of the reach of children. I should think so.

When I put the bottle back, I could swear I smelled chlorine on my hands. Did the stuff somehow seep through the plastic bottle?

Katherine looked at me. "I wonder with some of these products if you're *supposed* to smell what's inside them, and others you're not," she asked. It was a good question. Clearly, the smell of chlorine has come to represent a version of "clean," though not exactly the version of which Albert Donnay would approve.

For most people, chlorine conjures images of summertime pools and white whites and squeaky-clean bathrooms. And it is widely used for everything from water treatment to the production of bright white toilet paper and synthetic agricultural fertilizers. But it also has other roles: chlorine has been used for a hundred years as a chemical weapon—deployed against humans, not just bacteria. (Germany used it to horrific effect during World War I; it killed soldiers by burning out their lungs.)

Even before the jug of bleach reaches your home, chlorine production releases dioxins into the air, which can then enter the food chain—through cow's milk, for example. Chlorine factories are major emitters of mercury, which ends up in fish. Once it has been manufactured, chlorine is so toxic that just transporting it from one place to another becomes a major security risk. About 1.8 million carloads of hazardous chemicals move along the country's railways every year; some 100,000 of these loads—mostly chlorine and anhydrous ammonia—are especially dangerous. A cloud of gas released from a chemical train wreck is not something any mayor wants to see.

In July 2001, just two months before the terrorist attacks of September 11, a train carrying the solvent tripropylene derailed inside a Baltimore tunnel and shut down the entire city center. Rail lines still pass within feet of the stadiums where both the Baltimore Orioles and the Ravens play; when President Bush attended an Army-Navy football game in the football stadium, chemical shipments had to be suspended.

If you accidentally mix chlorine with ammonia, it can create lethal vapors called chloramines. But you knew that, right? You are warned right there on the bottle.

So a question: Does it strike you as odd that we've become so comfortable with a substance that Saddam Hussein used to make chemical weapons, a substance that is so toxic that just moving it around endangers entire cities? Just so you can have clean floors and white toilet paper? It's not like bleaching toilet paper with chlorine makes it any safer. For most products, chlorine bleach is purely for aesthetics; Americans have grown accustomed to white paper, so this is what the industry provides. In Europe, chlorine use is restricted, the toilet paper is typically brown, and nobody seems the worse for it.

You have doubts about these things. How could you not? Common sense may tell you one thing—that chemicals we once used as weapons might best be left outside the home—but saturation marketing has a way of changing your mind. Much of this marketing, not surprisingly, has taken its cues from the tobacco industry, which—despite decades of dire scientific warnings about the hazards—has managed to convince people that there is still some doubt about the dangers of smoking.

And so it goes with the chemical industry. One narrative, cultivated by industry, is that consumer products and the chemicals of which they are composed are safe. Consumers need not worry whether the products have been tested. Companies have the health and safety of consumers foremost in mind. Besides, they say, toxic chemicals are also found in nature. Lead is natural. So is salt. In small doses, both make life better. And it's not as if chemical companies need to convince you that a compound is safe for you to use it. All they need to do is make you doubt whether it is unsafe, and you'll keep on buying. "The whole thing is basically a front," said Mike Belliveau of Maine's Environmental Health Strategy Center, which directed the body burden study there. "They're seeking to manufacture uncertainty."

Our compulsive reliance on synthetic chemicals has insulated chemical companies from the public scrutiny one might expect from epidemics of mysterious environmental cancers and other illnesses. Belching DuPont plants dot the landscape near the University of Delaware, where I work. Driving by them is like driving by a slaughterhouse: you have a vague idea of what goes on in there, but beyond that, you'd rather live with an uncomfortable ignorance. In 2005, DuPont agreed to pay the EPA $16.5 million—including the largest administrative penalty the EPA has ever imposed—for covering up the health effects of the chemicals used to make Teflon at a West Virginia plant. Four years earlier, the company paid $107 million to settle a class action lawsuit brought by residents who claimed the company contaminated local waterways; the Teflon chemicals, which have been linked to both cancer and birth defects, were found in blood samples from both pregnant women and their fetuses. Yet the DuPont name graces a number of elegant buildings on my campus. Which impression, do you suppose, will my students remember?

Yet as consumers get more interested in buying "green" products, companies shift their marketing to take advantage of it. This "greenwashing" strategy only muddies the water further, Mike Belliveau told me.

"The people who make paintbrushes with plastic handles say, 'Buy our brushes because no trees were cut down,'" he said. "The people who make wood-handled paintbrushes say, 'Buy our brushes because our brushes aren't made with plastic.' At this point, Home Depot could pretty much call every product in their stores 'green.' This doesn't hold up to the scrutiny even of common sense."

A competing narrative is that industry has always had a financial interest in not testing its products adequately, in not disclosing all that it knows, and, when it is presented with evidence of possible toxicity, of fighting the accusations in court and the media rather than changing its formulations. Industry has long known the value of hiring its own scientists, and reaching into its deep pockets to fund studies that reflect favorably on its products. Independent research, funded in large part by federal grants, tends to ebb and flow, and an administration's ties with industry can greatly impede the competing narrative. With less funding for independent laboratory studies, there are few opportunities for a competing narrative to gain traction. This is a strategy Big Tobacco mastered and Big Chemical has taken to heart.

"But body burden studies are incredibly powerful and effective," Belliveau notes "Industries are very threatened by the idea that pollution is now in people. It used to be that the public thought that pollution was in the air, or in the water. But now they're starting to realize that pollution is in us."

So where do we turn?

In the home-cleaning aisles of some big box stores, a few mainstream companies are beginning to offer less toxic products. In the store we visited, we found two bottles of Clorox Green Works products, but we had to look hard for them. Theirs hardly seemed like a thriving market, at least under this roof. But their very presence was evidence of a shift. Just as mainstream supermarkets (and the industrial food makers that supply them) have responded to the demand for organic foods, so have mainstream big box stores begun to stock less toxic alternatives. Pick up a bottle of

Seventh Generation 2X Ultra Natural Laundry Detergent, and here's what you'll find on the front label: First, a big pink square at the top of the bottle that says: "You have the right to know." A whole different marketing approach, right up front. I could tell just by looking at the bottle that I was about to get educated.

The back of the label features the Seventh Generation credo, taken from the Great Law of the Iroquois Confederacy: "In our every deliberation, we must consider the impact of our decisions on the next seven generations." Immediately beneath that, you'll find a note from Jeffrey Hollender, Seventh Generation's "President and Chief Inspired Protagonist." At Seventh Generation, he says, "We Disclose All Ingredients."

The label presents a box chart showcasing the plant-derived cleaning agents ("low-foaming, stain removing power") and the non-animal derived enzymes ("premium-performance protein and starch stain removers [blood, grass, wine]").

Blood. Grass. Wine. Presumably, the typical user of Seventh Generation detergent would be sullying their picnic blanket by pricking their thumb on a corkscrew. But the marketing idea—selling products based on how nontoxic they are, in packages that go overboard in their listing of ingredients—made me curious. I decided to talk to the man who designs them.

Martin Wolf, a soft-spoken, middle-aged man with light blue eyes and a trim white beard, is the director of product and environmental technology for Seventh Generation, the Vermont-based company that has been a leader in making and selling nontoxic products for twenty years. A chemist, Wolf began his career working for the chemical giant eventually known as Ciba-Geigy. Later, as a consultant for the EPA, he got a firsthand look at the toxic by-products of manufactures from paper mills to electronics factories, and the incredible trouble citizens and governments have cleaning up industrial waste. He worked on notorious cleanup sites like Love Canal, and the Woburn, Massachusetts, water-poisoning site that became the basis for the bestselling book *A Civil Action*. Things have been a bit less dramatic since Wolf arrived at Seventh Generation, in 1990. These days, he devotes himself to designing things like liquid detergents, dish soap, all-purpose sprays, and glass cleaners.

But the work, in its way, is equally quixotic.

Americans spent over $432 million on all-purpose cleaners in 2007. Forty percent of that market was owned by three Clorox products: Pine-Sol, Clorox Clean-Up, and Formula 409. Seventh Generation has just 0.3 percent. But the ground is beginning to shift, and both Wolf and his counterparts at Clorox know it. Sales in the $2.7 billion household cleaning products market have been flat of late, but sales of "natural" cleaning products have been growing at 23 percent a year. Wolf bet years ago that the more people learn about what's in the products they buy, the better his company will do.

"Most American consumers don't make the connection between their shopping habits and the environment," Wolf told me. "But most people are concerned with their health, and the health of their family. Studies have shown that when women become pregnant, they get much more concerned about these things, things that could affect the health of their babies. Right now, most Americans say their grandmother used Tide, their mother used Tide. They say, 'That's what I'll use.' But that's changing."

When people first start thinking about reducing their exposure to toxic chemicals, they typically begin by thinking about food. They start looking at labels, noticing whether a vegetable is organic or a processed food has been made with trans fats. People may not keel over from eating a single head of lettuce sprayed with toxic chemicals, Wolf said, but that hasn't stopped organic food from becoming a bona fide "movement."

People have been much slower to recognize the connection between their health and the products they use in their homes. Phthalates, for example, are not only used in cosmetics and plastic bottles, they often serve as solvents that keep fragrances suspended in things like laundry detergent. Butoxyethanol, a cleaning compound found in engine degreasers, is also an ingredient in conventional household cleaners like Formula 409, Fantastik, and Windex. (In its Household Products Database, the National

Institutes of Health reports that the compound has been shown to cause problems—in lab rats, at least—with the central nervous system as well as in the kidneys and liver.)

"People eat four to five pounds of food a day, but they breathe twenty pounds of air a day," Wolf said. "If you are active, you breathe forty pounds of air a day. Who thinks about Formula 409 having butoxyenthanol, and thinks about the effect it has? It's cheap. Who thinks about synthetic fragrances on their clothing? Conventional manufacturers want you to smell their product when you walk down the aisle in the supermarket, they want you to smell it when you put it in the wash, they want you to smell it when you put on your shirt. These companies spent a lot of money on research getting these fragrances to be recognizable. But fragrances are volatile. There is a lot of chemistry in this."

The interplay between body chemistry and environmental toxins is so complex that, especially once you account for the debate over low-dose exposure rates, "proving" that an individual product is dangerous will always be elusive. Martin Wolf knows this. So do the people running Big Chemical, who learned it from Big Tobacco.

"Do we really know how low that risk is? The experiments still haven't been done. And they never will be, frankly," Wolf told me. "A really good carcinogenic study is done with fifty rats. So fifty rats represent three hundred million people. That means each rat represents six million people. If three rats out of fifty get sick, is that a significant deviation? If you're working with fifty rats, those three sick rats represent eighteen million people.

"We have three hundred million people in this country. Twenty-five to thirty percent are smokers. So you're looking at seventy-five to ninety million 'rats' in your control group, and we still can't 'prove' smoking causes cancer. Our systems are so complex. You just can't isolate toxins like that. And you never will. The quest for a smoking gun is a hopeless quest. And industry will always hide behind that fact."

When he begins designing a new product, Wolf will consult with chemists at the company's manufacturing plants. "I'll say, 'Here's a formula for an existing product. We want to find substitutes for this, and this, and that," Wolf told me. The chemists will test them—though never on animals, a company policy—and see whether or not they work. It's a process that moves up from the company's philosophy, rather than a process driven only by a goal of maximum chemical effectiveness.

Wolf acknowledges that designing products without being able to rely on toxic compounds restricts his chemical palette. Instead of phosphorous, for example, which costs 15 cents a pound, Seventh Generation uses sodium citrate, which costs 40 cents a pound—a significant expense when you're talking about 20 percent of a product. "Industry always brings the same arguments," Wolf said. "'It can't be done,' they say. 'There's not enough of the bad stuff in there to make a difference.' When companies say it can't be done, what they really mean is that it can't be done and still maintain the same profit margins. We say you can make products that are equally effective. We say the standard products are hazardous."

Take chlorine. Seventh Generation uses no chlorine in the pulp for its diapers, feminine hygiene products, or toilet paper, a practice in line with standards in a number of European countries, but not in the United States.

Chlorine seems to be something of a bête noire for Martin Wolf. Before the late 1990s, paper mills produced 15 pounds of chlorinated hydrocarbons for every ton of pulp, Wolf told me. If a company made 200 tons of paper per day, that also produced 3,000 pounds of chlorine a day, most of it discharged into streams or as solid waste. This toxic waste was in addition to the other common wastes from paper production, like dioxins and PCBs.

In 1998, over the loud objections of industry, the EPA passed a "cluster" rule setting contaminant levels from chlorinated products. This quickly reduced the amount of the "absorbable organic halogens," or AOX, by "almost an order of magnitude," Wolf said.

Europe and Latin America do far better at using recycled pulp for their toilet paper: 20 percent versus 2 percent in the United States. The impact of such a difference is vivid, and not only in the volume of chlorine we produce: Greenpeace claims that Kimberly-Clark, which makes the Cottonelle and Scott lines of tissue, gets nearly a quarter of its pulp from two-hundred-year-old trees in Canada's old-growth forests.

"There is something absurd about cutting down two-hundred-year-old trees for a three-second application," Wolf said. "Even cutting down eighty-year-old trees from a tree farm for this is absurd. Recycled paper is the only sensible thing." ("Recycled toilet paper" is a phrase that would raise the eyebrows of any eight-year-old, and Wolf is quick to point out that in this case, the recycling happens before the paper is used, not after. In Seventh Generation's case, 20 percent comes from recycled magazines.)

As companies like Seventh Generation—and Method, Mrs. Meyer's, and Ecover—have made their way into the mainstream—and a bigger share of the market—much larger corporations have responded by refining their formulas, or by buying the competition. SC Johnson boasts that its Shout and Scrubbing Bubbles products have been cleared by the EPA's Design for the Environment program; for good measure, it also bought Mrs. Meyer's, another leading manufacturer of cleaning products marketed as nontoxic. Colgate-Palmolive introduced its Palmolive Eco dishwashing detergent and claims that it's the first mass-marketed automatic dishwashing detergent brand to eliminate phosphates—and it bought Tom's of Maine. Clorox has added the Green Works line to its $4.8 billion family of household cleaning products—and it bought Burt's Bees.

Martin Wolf is unmoved by such alliances. He points out that Procter & Gamble still takes its cue straight from the Big Chemical playbook (claiming more research is needed) and that the "green" Clorox line still uses chlorine in its cleaning products, butoxyethanol in its spray cleaners, and petrochemical based surfactants in its detergents. He also knocks Clorox for listing the ingredients on its Green Works cleaners, but not on its regular products. "Clorox is in the position of having told people for a hundred years that their products have been as safe as they can possibly be, and now they're telling people they have a new 'safe' product," Wolf said. "If you confuse people, they become incapable of action."

To Martin Wolf, this rhetoric leaves a hole in the marketplace big enough to drive a Seventh Generation truck through—a hole made wider recently by progressive legislation. Since September 2006, a New York State law has required schools to use cleaning products that do not contain carcinogens, reproductive toxins, or scents that could aggravate asthma. Other states are also encouraging changes in the way public buildings are cleaned. For years, Wolf lobbied California legislators to pass a pair of bills to regulate the use of household chemicals and create a database so consumers can find more information on what goes into them. In September 2006, the state legislature passed both. Two years later, the state passed a law requiring dry cleaners to stop using older perc machines by 2010, and to stop using the chemical altogether by 2023. California has also established a list of chemicals next up for regulation, based on how ubiquitous they are in the environment and on their potential impacts on infants and children.

In the last couple of years, Seventh Generation has begun making concerted—and not uncontroversial—inroads into mainstream supermarkets. In January 2007, Stop & Shop added the company's products to three hundred supermarkets from New Hampshire to New Jersey. Then, eighteen months later, the unthinkable happened: a deal with the mother of all big box stores, Walmart. Walmart launched a "retail concept" called Marketside, initially in four small stores in Arizona, which now sell prepared meals, natural and organic food—and a line of cleaning and paper products from Seventh Generation. It's not as if Walmart makes such moves out of the goodness of its corporate heart. It has seen a market for nontoxic products, and it has responded by providing them.

"I've said that hell would freeze over before Seventh Generation would ever do business with Walmart," said Seventh Generation's Jeffrey Hollender after the deal was announced. "Now I've got to concede that I was wrong."

FOUR The Tap

TOXICS IN THE TAP WATER

You know that old cliché about Zen masters who can look at a drop of water and see the whole universe? Jerry Kauffman is kind of like that. When he looks at a glass of water, he sees entire systems at play. He knows, with a striking degree of precision, exactly where his water has come from. Not just from which pipe, or from which treatment plant, or even from which river; he knows from which contaminated farm fields, and which shopping mall parking lots, and which sewage treatment plants upstream. If the water is slightly brown, he knows there's iron leaking from old municipal pipes. If it's winter and the water has a slightly yellowish-blue cast, he knows cities have been spreading deicing chemicals on the roads.

Other things Kauffman can tell with his tongue. He can taste when there's too much chlorine in the water, or not enough. If it's summer and the water smells like smoke, he knows there's been another compost fire at an upstream mushroom farm and the ashes have washed into the creek. If it's winter and the water smells a little like cucumbers, it means towns upstream have been salting their roads with urea—yes, you read that right—that they get from fertilizer makers.

Other things he can't see or smell, but just knows intuitively. He knows, for example, that there are 700,000 chickens in his watershed, many of them on farms that do little to control the enormous amount of waste flowing into local streams. He knows there are industrial chemicals (like benzene) and naturally occurring toxins (like arsenic) that are part of the region's soil. He knows that his water, like all of our water, contains pesticides, petrochemicals, and a vast array of prescription drugs. Mood enhancers. Antipsychotics. Erection stimulators. Birth control hormones. Anticonvulsants. All the waste, in other words, that flows downstream from everyone living upstream.

One recent afternoon, I handed Jerry Kauffman a glass of tap water and asked him to tell me about it. We were sitting in a university dining hall, and I had drawn the water from a nearby drinking fountain. The fountain, situated alongside a well-traveled corridor, probably served hundreds of students a day, and the distribution pipes, somewhere beneath our feet, served many thousands more. Where the water had come from, exactly, was something of a mystery, as it is in most urban and suburban places. There were no creeks or streams visible anywhere on campus or in town; they had long since been buried underground. The nearest river was a mile away, in the lone patch of woods that had somehow escaped the bulldozers.

I had been drinking this water for years, without ever thinking very hard about it, and now, sitting in the dining hall, staring into this glass, it looked clear enough to me. But the truth is, when it comes to water, I don't have Jerry Kauffman's eyes. When I handed him my glass of water, Kauffman swirled it around, like Robert Parker scrutinizing a fine Burgundy. He poked his nose into the glass, then held it up to the light. "Water is supposed to be clear and odorless," Kauffman said. "If you can smell something, usually it's bad for you." He sniffed again. "You can taste it if there's too much iron in your water, or too much chlorine. Some of these chemicals have an odor threshold, and some of these we can test for. What I don't know is which of the other thousands of chemicals are in here that we haven't tested for."

Kauffman is Delaware's state water coordinator, which means it's his job to make sure that when Delaware citizens turn on their taps, they aren't being slowly poisoned by the myriad chemicals flowing into their drinking water. This has become quite a challenging job in recent years, especially since water pollution is no longer the blatant, visible travesty it was back in the 1970s, when Americans somehow got used to their urban rivers catching on fire. It's far more subtle now. Not less ubiquitous. Just more subtle.

Kauffman grew up in Pennsauken, New Jersey, right across the river from Philadelphia, in a town so congested he could run the three miles to school faster than he could take the bus. All that exercise got him in shape, and by the time he graduated and joined the track team at Rutgers, he was running eighty miles a week. He got his time in the mile down to 4:16, and came in second in his conference in the 10,000 meters. Thirty years later, he still has his old white track suit, still stained red from the Brunswick shale dust he kicked up training along the Raritan River.

In the winter, when it got too cold to run, Kauffman and his friends would strap on skates and play hockey out on the frozen Delaware River. It was there that he got his first glimpse of just how polluted his native waterways had become. "The Delaware during the 1970s was black," Kauffman said. "It was a dead river, black with the crud from oil spills, black with raw sewage. The ice was black, the shoreline was black. In summers, we'd go bow fishing, and the carp would be coated in the stuff. Back then, I just assumed that all rivers were black."

Kauffman is the only man I know who speaks wistfully about the air quality in northern New Jersey. He remembers the air around Philadelphia, especially that blowing over from chemical plants in Port Richmond, being so contaminated that it actually smelled sweet—and not in the nice sense of that word. It wasn't until he drove a couple hours north to Rutgers that he smelled clean air. (A few dozen miles north of that, he notes, you were in Newark, and the air started smelling sweet again.)

At Rutgers, Kauffman began as a history major but then switched to engineering so he could spend more time working outside. In the summer, he got a job at the Jersey Shore taking soundings for a bridge project. What he remembers most vividly from that time was the sight of blue crabs floating upside down, dead from pollution.

"It made me really angry to see that," Kauffman said. "That's when I knew I was going to study water."

When I asked him to tell me more about the safety of that glass of tap water in his hands—to explain how things have gotten so contaminated and what can be done to protect ourselves—he suggested we take a ride in a canoe. You can't understand what comes out of your tap, he said, without understanding where that water has been before it gets there. And where that water has been before it comes out of your faucet might surprise you.

Kauffman and I made a plan to launch a boat on the Brandywine River, which provides much of the drinking water for the city of Wilmington. But really, he could have suggested any river serving any population in any city in the country: the Chattahoochee in Atlanta; the Gunpowder in Baltimore; the Hudson in New York; the Schuylkill in Philadelphia. I have paddled all of these rivers, and many others, and they all share the Brandywine's troubles. Millions of people live, work, and drive in these rivers' watersheds. They build homes and factories and vast parking lots and huge crop and livestock operations near the river-banks, and the rivers absorb, one way or another, every ounce of in-fluence humans place on them. Yet these same people also demand that their rivers and reservoirs supply them with millions of gallons of fresh, clean drinking water every day of the year. That's a lot of stress for any watershed to bear.

For generations, Americans have treated our river systems like sewer pipes, flushing our toilets, our factories, and our farms to a distant place few of us ever see. The trouble, of course, is that there is no "other place." Water, as any seventh-grade science student knows, has a way of cycling—endlessly—through both our earthly and our bodily systems. There is the same amount of water on the earth today as there was ten thousand years ago. It's just more polluted.

Although thousands of synthetic toxins have been studied by government and independent scientists, not a single chemical has been added to the Safe Drinking Water Act since 2000. Many drinking water safety standards have not been updated since the 1980s; others have not changed since the law was passed, in 1974. By one estimate, in the last five years, some 62 million Americans have been exposed to chemicals that fail at least one government safety standard. The EPA has only recently vowed to begin analyzing chemicals known to be endocrine disruptors—more than thirteen years after Congress first called for it.

"Surprisingly little is known about the extent of environmental occurrence, transport, and ultimate fate of many synthetic organic chemicals after their intended use," the U.S. Geological Survey has reported. Among the primary concerns: hormonally active chemicals, personal care products, and pharmaceuticals that are designed to stimulate a physiological response in humans, but also affect plants and animals. Compounds in these products are now thought to impair reproductive organs, increase the incidence of cancer, and add to the toxicity of other chemicals already in the water supply.

Some 97 percent of the nation's rivers are contaminated with at least one pesticide. Even compounds like DDT, chlordane, aldrin, and dieldrin—most of which have been outlawed for decades—are still being found in fish and riverbeds. More than 80 percent of urban streams and nearly 60 percent of agricultural streams are contaminated at levels dangerous to aquatic life. A single pesticide, atrazine, has been found in 75 percent of stream samples and 40 percent of groundwater. New research suggests that even at concentrations meeting current federal standards, atrazine may be associated with birth defects, low birth weights, and menstrual problems. Or take PCBs. Like countless other industrial chemicals that resist decay, PCBs gradually make their way up the food chain, from water-borne phytoplankton to larger and larger fish, then to birds and to humans. Scientists have found that PCB concentrations in animal tissue can be *25 million times* their concentrations in the surrounding water.

So that's what's happening nationwide. In Delaware, it's Jerry's Kauffman's job to make sure his river's health doesn't collapse. And to try to convince people that a river's health is directly tied to our own.

UPSTREAM

One lovely afternoon in late May, I met Kauffman and a group of his graduate students in the rear of a vast mall parking lot in suburban Wilmington, where a small river outfitter had set up shop. We tried on life jackets, threw a bunch of paddles into the back of a university van, and lashed our boats to the roof. As we made our way out of the parking lot and maneuvered through a long strip of fast food restaurants and chain stores, Kauffman looked out the window and reminded me that whatever rain hit the pavement under our wheels would eventually make its way into the Brandywine. In a watershed, he said, everything drains down to the river.

Everything.

Fifteen minutes later, after navigating suburban traffic, we rolled down a hill near Chadds Ford, Pennsylvania, and pulled into the parking lot of the Brandywine River Museum.

The Brandywine Valley is particularly rich in history, and not least for its place in the history of chemistry. Settled some 350 years ago, the valley became an important center of commerce in the British colonies; then, in 1802, the Brandywine caught the eye of a French immigrant named Eleuthère Irénée du Pont de Nemours. Du Pont paid a bit over $6,000 for a parcel of land along the riverbank to build a mill so he could make gunpowder; his company soon became the largest manufacturer of gunpowder in the country. Over the next two hundred years, of course, DuPont would grow into one of the wealthiest and most influential companies in the world, stitching dozens of products deeply into our lives: polyester, nylon, Teflon, Lycra, neoprene, Mylar, Tyvek,

Kevlar. It wasn't lost on me that the plastic that had gone into making our canoe had almost certainly been a DuPont product.

I helped Jerry Kauffman and his students unstrap the boats. We propped them on our shoulders and made our way down a dirt path to the river. Kauffman and I put our boat in last; we wanted to make sure the students began their journeys on top of the water, and not beneath it.

Which is a problem when you're out on a river on a sunny day in May, especially if you're a graduate student studying watershed management. It wasn't ten minutes before Kauffman's students were splashing one another with their paddles, jumping out of their boats, and generally behaving like, well, like anyone should on a beautiful day on a beautiful river.

The Brandywine drains 320 square miles in southeastern Pennsylvania and northern Delaware, and even from our canoe, it was quickly apparent why it wasn't only industrialists who had come to love the valley. The river fostered the Brandywine school of painters—Howard Pyle, N. C. Wyeth, Andrew and Jamie Wyeth—and today, with its lovely farms, grand DuPont estates, and rolling fields, fox hunting and polo still maintain a hold on the region's imagination. The valley is home to Winterthur, the grand museum devoted to Henry Francis DuPont's collection of antiques and Americana, and Longwood Gardens, a thousand acres of manicured gardens developed by Pierre S. DuPont. This is pastoral country, a quilt of horse, cattle, and pig farms, corn and soybean farms, mushroom farms, and apple orchards.

To my eyes, the water looked clear enough—but then again, so had the glass of water I'd offered Kauffman back in the cafeteria. The forested banks on either side seemed remarkably intact for a river running though one of the most densely populated regions of the country. But the truth was more complicated. Long before the Brandywine passes into downtown Wilmington, and well upstream of the DuPont estates, the river has already been highly contaminated, not once but sixty times by wastewater discharges—from single homes to hospitals to the municipal water treatment plants in Coatesville, Downingtown, and West Chester,

Pennsylvania, which collectively treat and discharge 10 to 15 million gallons of water every day. To put this number in perspective, consider that during a drought the Brandywine flows at about 21 million gallons a day—which means that during stretches in the summer, more than half the water in the river is treated wastewater.

The river also collects vast quantities of chemical runoff from farms in southeastern Pennsylvania. It absorbs pharmaceutical drugs flushed through untold thousands of septic tanks and subdivision sewer lines. It is buffeted by industry, and the contaminants washing off countless square miles of roads and parking lots. Even the smaller-scale animals on farms in Pennsylvania's Amish country produce great quantities of nitrogen. And if it rains in Amish country on Sunday, the runoff is in Wilmington by Monday, and people in Delaware are drinking the treated water on Wednesday.

Yet just how polluted you consider the river to be depends on your sense of history. "Compared to presettlement, the Brandywine is nowhere close to being as clean as it once was, but compared to a hundred years ago, it's recovering," Kauffman told me. "People used to know the type of ink being printed at the paper plant by the color of the stream."

Water here—as everywhere on earth—has always created conflict. The words "river" and "rival," after all, come from the same root. If companies can get someone else—someone downstream—to deal with pollution, they tend to do it. "That's what pollution is all about," Kauffman said. "Someone else is always left to deal with it. Companies just build pollution fines into their bottom lines, into their cost of doing business. Industry calls pollution an 'externality,' something they don't have to deal with. That's not what I call it."

Kauffman spends a good bit of his time reminding people that the water we drink is the same water that flows through heavy industrial plants, drains farm fields, and crosses state lines. In certain parts of Philadelphia's Wissahickon Creek, whose water flows into the Schuylkill and then into the Delaware River, more than 20 percent of the fish have diseases, tumors, or fin damage.

This an especially difficult lesson to impart in states (like Delaware, sandwiched as it is between Pennsylvania and Maryland) that get most of their drinking water from rivers pouring in from other places. What this means is that Kauffman constantly has to remind Pennsylvania lawmakers that their wastewater ends up in Delaware's drinking supply. It also means that when he goes over to the state capital in Dover, he has to convince lawmakers to spend Delaware money planting trees across the border in Pennsylvania. Thick, forested buffers prevent all kinds of waste products from flowing straight into rivers, and planting trees is far cheaper than retrofitting water treatment plants.

"Water treatment systems are as good as they have ever been in human history," Kauffman told me. "But water protection is also incredibly underfinanced in the United States. If you total up all the water quality programs in the country—all the testing, all the watershed protection efforts—it comes out to about fifteen bucks a year for every man, woman, and child. That's the level of commitment to clean water. Is that enough? I say it's not. Compare that to the other pieces of the federal budget pie, and the water quality piece is so small you can't even see it on a pie chart."

Before the federal Clean Water Act was passed, in 1972, people and industries alike could pretty well dump whatever they wanted right into the river. I remember paddling a river in Georgia in the early 1990s with a water quality expert; he pointed to a spot that had been used, a couple of decades ago, as a dump for a chicken processor. Up and down the stream, little kids had swum in a river choked with chicken parts. No longer. Industries treat some wastewater on-site; municipal plants do the best they can with the rest. Yet over the last five years alone, industries have broken water pollution laws more than half a million times—by doing everything from not reporting their emissions to dumping chemicals that may cause cancer or birth defects. Only 3 percent of these violations resulted in fines.

You might think that big rains would help dilute all the toxins and bacteria in a river system, but that turns out not to be the case.

Heavy rains can bring massive nitrogen bursts from fertilizers running off farms and lawns; where a normal (and "safe") nitrogen level might be 10 parts per million, a post-rain reading might be 500. "It used to be thought that 'the solution to pollution was dilution,' " Kauffman said. "Lakes and rivers were thought to be big enough to dilute the sewage."

We now know that's not true: contaminants don't disappear just because they cross state lines. "People in Dover always ask, 'Why is my money being spent in Pennsylvania?' " Kauffman said. "They don't understand that that's where their water comes from."

A mile or two into the trip, Kauffman and I passed several families picnicking and swimming along the riverbank. The families looked to be from Mexico or Central America. Didn't they realize this was America, I thought? Didn't they realize that in this country, we consider our rivers to be too polluted to enjoy? Didn't they realize that all those lovely farm fields, to say nothing of the subdivisions that lay beyond, were also the source of enormous quantities of waste?

To be fair, it's not like they could see the sources of the pollution. Where public health experts a hundred years ago worried about water-borne illnesses like typhoid and cholera, they now fret over pharmaceutical drugs, pesticides, and the mountains of manure (and attendant gastrointestinal bugs) pouring off farms. As Chris Crockett, director of planning and research for Philadelphia's water department, put it to me, water quality is directly affected by what comes out of a cow's butt. In the Brandywine Valley, this is no joke: a recent census listed Chester County, Pennsylvania, as having some 42,000 cattle, 13,000 hogs, 8,600 horses, 3,000 sheep, and nearly 700,000 chickens—double the number of human residents.

Animal waste wreaks havoc on drinking water, not only because of the nitrogen and phosphorous it contributes, but because of the antibiotics and the microorganisms flushing through the animals' guts. It works like this: farm animals eat, they poop, it rains, and the waste runs into the nearest creek. From there it's into the river, into the drinking water supply, and into your glass. Recent studies have found the fecal bacteria giardia in 90 percent of samples taken from the Brandywine. Cryptosporidium, another parasite, is common in the guts of ruminants like sheep and goats, and passes easily into the guts of humans.

It's not just drinking contaminated water that can make you sick, of course. Swimmers can become contaminated, and so can fishermen. Baltimore, where I live, sits at the lip of the Chesapeake Bay, an ecological bowl that collects the waters of the middle third of the eastern United States. Water pours into the bay from as far away as Cooperstown, New York, and the Blue Ridge Mountains of Virginia, a stretch that includes enormous swaths of agricultural land and unimaginable quantities of animal waste.

A few years ago, Johns Hopkins's Ellen Silbergeld and her colleagues Jennifer Roberts and Thaddeus Graczyk found that a lake near my house was so contaminated with cryptosporidium that a fisherman's chances of becoming infected were 80 percent. Not by eating his catch, mind you, but by merely touching the water and then putting his hand to his eyes or mouth.

Now, you might say, I'm not crazy enough to fish in an urban stream; streams aren't clean enough to fish in. Maybe so. But in 1993, more than 400,000 people in Milwaukee fell ill, and more than 100 died, because of an outbreak of cryptosporidium. Gary Wells, a Milwaukee resident and an AIDS patient, told CNN that he lost more than one friend to the parasite. "I used to drink this," he said, gesturing toward his sink, "because I thought I could trust that it was OK. But obviously I was wrong."

In the spring of 2008, almost forty years after Cleveland's Cuyahoga River caught fire, sending flames eight stories into the air, Dr. Silbergeld and her Johns Hopkins colleague Jay Graham published an essay called "The Cuyahoga Is Still Burning." It's burning not with fire but with viruses, bacteria, and microparasites. Fish and people might be returning to swim in the river, the scientists wrote, but the ever-expanding volume of contaminants running off industrial farms and suburban sprawl is too much for any system to handle.

The problem is not just animal waste, either. If rainfall (or snowmelt) is heavy enough, a city's sewage treatment plant can be overwhelmed by water gushing in from the countless pipes and storm drains in the streets. Before some recent upgrades to its sewage treatment system, Wilmington's system overflowed 60 percent of the time it rained. Nationwide, the EPA says, this happens about 40,000 times a year. Some researchers estimate that more than 860 billion gallons of untreated sewage gets into our waterways every year.

In Baltimore, sewage treatment systems are so decrepit that in 2002 some 355 million gallons of raw sewage flowed into city streams. This was more than ten times the previous year's amount. Under pressure from the EPA and the Justice Department, the city agreed to spend nearly $1 billion to upgrade its system by 2016.

And while the health implications of raw human and animal sewage are hard to tally, Jerry Kauffman has, shall we say, a gut feeling.

"It's hard to quantify spikes in ER visits, but they're out there," he told me. Nearly 20 million Americans get sick every year from water contaminated with parasites, bacteria, or viruses, a number that does not include illnesses caused by synthetic chemicals.

And what is true in miniature in a place like the Brandywine Valley is true writ large elsewhere in the country. Small farms that once pastured their cows and considered manure a priceless fertilizer have given way to large farms that keep their animals in cement-floored barns. Rather than permitting cows to do what they do best—spread manure on fields—farmers now use front-end loaders to move the waste into vast containment ponds, which can leak or overflow altogether. Farmers on the eastern shore of Delaware, Maryland, and Virginia raise some 570 million chickens, which collectively produce some 1.5 billion pounds of manure a year. This is more than the total human waste produced by New York City, Washington, D.C., San Francisco, and Atlanta combined, yet it essentially flows untreated from farm to river to bay, producing 42 percent of the nitrogen and 46 percent of the phosphorous pouring into the Chesapeake.

Paddling along the Brandywine, Jerry Kauffman reeled off some of his favorite water-quality statistics. By the time the Colorado River reaches household taps in southern California, more than two hundred communities, including Las Vegas, have discharged their effluent into the river. By the time someone in New Orleans drinks a glass of tap water from the Mississippi River, Kauffman says, the water has traveled through the guts of five people. How many cows have also passed the water, and how many petrochemical plants, Kauffman did not hazard a guess.

The trouble is that rivers, like air currents, don't adhere to state lines. As a history major–turned–engineer, Kauffman is fond of invoking Benjamin Franklin, whom he calls "the founder of America's first municipal water system." Franklin left money in his will to purchase land in the Schuylkill watershed to protect Philadelphia's water. He also petitioned Pennsylvania to move its tanneries and slaughterhouses away from the shores of the Delaware River.

If you really want to clean up a natural system, Kauffman says, you might need to reimagine political boundaries. If the country were divided up into thirty-six watersheds instead of fifty states, people would likely live with far clearer connections to their drinking water—and might take more care in protecting it. And then people like him wouldn't have to work so hard to get a Delaware legislator to pony up for clean water in Pennsylvania, because both areas would be drinking the same water, and both would know it. (Such a system has long been in place in Switzerland, where political boundaries are dictated by ridgelines, which determine the way water flows.)

But given American history, such an arrangement does not seem likely. A century ago, Kauffman said, John Wesley Powell, the famous explorer of the Colorado River and head of the U.S. Geological Survey, proposed dividing the American West along watershed lines. He lost his job for his trouble. More recently, the Supreme Court declared—in a ruling hailed by the developers who brought the case, as well as their allies in industry and Big Agriculture—that the Clean Water Act applies only to "navigable, permanent waterways." The rest of our fresh water—including 60 percent of the country's streams and some 20 million acres of wetlands—apparently do not need such protection.

Just a couple of weeks before Kauffman and I put our boat in the water, the Associated Press had reported that "a vast array of pharmaceuticals" had been found in the drinking water supplies of at least 41 million Americans. While the concentrations were measured in parts per billion or even trillion, far below those of a medical dose, they nonetheless raised real concerns. People drink a lot of water, after all, and they drink it every day. For many, many years. They also bathe in it. And cook with it. Our blood is 95 percent water. The bodies of babies are nearly 80 percent water; adult women and men are 50 to 60 percent water.

Over the course of a five-month investigation, AP reporters reviewed hundreds of scientific reports, analyzed federal drinking water databases, and interviewed 230 officials, academics, and scientists. They also surveyed the nation's fifty largest cities and a dozen other major water providers, as well as community suppliers in all fifty states. Their discoveries were alarming: drugs had been detected in the drinking water supplies of twenty-four major metropolitan areas, from southern California to northern New Jersey. Watersheds—the systems of creeks and rivers that both form and drain into our water supplies—were also found to be widely contaminated.

Officials in Philadelphia reported fifty-six pharmaceuticals or drug by-products, including medicines for pain, high cholesterol, epilepsy, mental illness, and heart problems. More than 18 million people in southern California drink water contaminated with antianxiety medications. Nearly 900,000 people in northern New Jersey are exposed to angina medications and the anti-seizure and mood stabilizer carbamazepine. San Franciscans get sex hormones. Water officials did their best to dampen anxieties. The public "doesn't know how to interpret the information," the head of a California water supplier said. We don't want to get people alarmed.

In the canoe, I brought up the story and asked Jerry Kauffman where all these drugs had come from. Was somebody dumping them illegally? Had there been an accidental spill? Nope, he said; the contamination was nationwide. The drugs came from inside us. It turns out that people's bodies do not fully absorb medications, and what is not absorbed ends up being excreted and flushed down the toilet. Although municipal wastewater is treated before it is piped back into rivers or lakes, and drinking water is treated before it is piped into our homes, not all drug residues are—or can be—extracted by the treatment process.

When I asked him about all the drugs in our drinking water, Kauffman smiled and offered the usual jokes about what a glass full of antidepressants might mean for the national mood. But then his demeanor grew more serious.

"How concerned should we be? Even the Bush White House," he noted, "issued a policy, asking people not to dump these things down the toilet, which is what we'd always been told to do. But these pharmaceuticals ought to be handled like all other chemicals. We should monitor for them and set standards. The drug companies should be regulated like any other industry that disposes of toxins. They make the stuff; they should have a hand in dealing with its disposal."

The trouble is, treating these contaminants is not as easy as dumping chlorine into the water treatment tanks. Try as they might, engineers simply cannot invent a system capable of stripping every chemical compound out of contaminated river water. Water treatment plants, many of them built in the late nineteenth century, were designed to prevent outbreaks of things like typhoid and cholera. The engineers who built them never imagined they'd need to remove things like Viagra.

The federal government does not require testing for pharmaceuticals in drinking water, and has not set safety limits for them. Of the 62 major water providers contacted by the AP, only 28 tested for drugs. Among the 34 that don't are Houston, Chicago, Miami, Baltimore, Phoenix, Boston, and New York's Department of Environmental Protection, which provides water to 9 million people.

This has become a troublesome fact, particularly as prescription drug use has exploded in recent years. Over the past five years, drug prescriptions in the United States have risen 12 percent, to 3.7 billion. Nonprescription drugs have remained steady at 3.7 billion. Veterinary drug prescriptions rose 8 percent, to $5.2 billion over the same period.

The industry response to the AP's report was boilerplate. "Based on what we now know," said Thomas White, a consultant for the Pharmaceutical Research and Manufacturers of America, "I would say we find that there's little or no risk from pharmaceuticals in the environment to human health."

But recent medical studies have shown that even small amounts of medications can affect human embryonic kidney and blood cells, as well as breast cancer cells. In rivers, male fish are being feminized. Even earthworms, which inhabit the very baseline of the food chain, are contaminated. A year before the AP report, the director of environmental technology for Merck and Co. had told a conference that "there's no doubt about it, pharmaceuticals are being detected in the environment and there is genuine concern that these compounds, in the small concentrations that they're at, could be causing impacts to human health or to aquatic organisms."

The AP's report was not the first time an alarm bell has gone off about the nation's drinking water supplies. More than a dozen years before, the U.S. Geological Survey had surveyed 139 streams around the country and found toxins left over from things like pharmaceuticals and cosmetics in 80 percent of them. Beyond providing further evidence of broad chemical contamination, the study confirmed what many water experts had feared: that toxic chemicals not only don't break down, they survive municipal wastewater treatment. And cities and states just have not been able to catch up to the new contaminants.

"This is purely a cost issue," Kauffman told me. "Better technology is achievable. We can screen for these chemicals, we can set safety standards for them, and there is technology out there to treat it. The costs for reverse osmosis and ultraviolet treatment are coming down, which could allow us to stop using chlorine. We could screen for medicines just like we screen for zinc or anything else. It's just a matter of cost."

But Kauffman does not spend his days trying to convince lawmakers to build more expensive treatment plants. "This is not an engineering problem, it's a social behavioral problem," he said. "It used to be that all the emphasis was on the tail end. Most engineering programs are still locked into a 'we can fix it with technology' mind-set. Now we are trying to emphasize prevention. But it's the big-picture behavior modification that I see as being the main challenge. We need to get people to identify with streams and change their behavior.

"The most cost effective way to protect a watershed is through source water protection. Cities like Seattle, San Francisco, and New York City do this very well—they get their water from watersheds that are 80 percent forested. Just increasing forest cover by 10 percent can dramatically increase the money you save on water treatment. It's much less costly to plant a forest than it is to build a new treatment plant."

Another enormous headache for water treatment experts like Jerry Kauffman is the runoff that pours off what are known as "impervious surfaces"—roads, parking lots, even the roofs of people's houses. Americans have paved 4 million linear miles of public roads, and this number does not include the 43,480 square miles of parking lots, driveways, or other paved surfaces. Road salt, petrochemicals, degraded brake linings—all of it ends up in rivers, and then in water treatment plants. In Tysons Corner, Virginia, forty years' worth of turbocharged suburban development has turned a sleepy section of farmland into an office and shopping district with 46 million square feet of buildings and 40 million square feet of parking lot. That's a lot of cement—and a lot of storm water runoff, all of which ends up in the Chesapeake Bay. In and around Seattle, the volume of petrochemicals flowing from roads and parking lots into Puget Sound is twice the volume spilled by the *Exxon Valdez*. These chemicals don't disappear, of course. They go right back into the food chain. People used to think of Puget Sound as a toilet, Washington's governor said, but really it's a bathtub.

In Delaware's tidal streams, PCBs, the jelly-like compounds used to insulate electrical transformers, have been an especially serious problem, Kauffman noted, not because they were dumped there (as they were by General Electric in New York's Hudson River) but because so many PCBs were used for things like the Amtrak rail lines. Time was, engineers built electrical substations right next to water-intake stations. "We wouldn't do that again," Kauffman said.

Someone has to pay to clean up all those toxins, and Kauffman and others think the money should come not from a state's general fund but from the people who paved things over in the first place. "Why should a guy with a small row house pay the same as a guy with a ten-thousand-square-foot parking lot?" Kauffman asked.

As it is, people don't think about water treatment, because they pay less for it than they do for cable television. Once people have to fork over more, the theory goes, they'll start to pay more attention. Kind of like driving a smaller car when gasoline goes to $4 a gallon.

Here's how it would work: Let's say you build a massive parking lot next to a shopping mall. You no longer get to watch the water flow down into storm drains and not think about it anymore. Same with the driveway at your house. On the other hand, if you can figure out a way to keep water on your own property— rain barrels hooked up to your downspouts, rain gardens designed into your lawns—you would get a tax credit. The setup would be similar to people installing solar panels on their roofs and selling electricity back into the grid. You take care of your own place, in other words, and you benefit. And by taking care of your own place, you remove your part of the burden from the larger system as a whole. Thought of this way, backyards could function not just

as lawns but as water purifiers: the rain that falls on your yard filters through your gardens and your soil, and thus doesn't flow off
into the neighborhood's storm sewer, where it doesn't flow into a
river, where it doesn't need massively expensive water treatment
plants. Now, to be sure, this is—dare I say it?—something of a pipe
dream, but you get the picture. The city of Philadelphia began
charging more for storm water treatment in 2010; it has also
pledged to spend $1.6 billion over twenty years to install rain gar-
dens, build porous sidewalks, and plant thousands of trees. Where
once businesses got two bills (one for water and one for sewer),
now they will get three.

The trouble is, this sort of thinking is new enough to seem
threatening. Government requests that farmers or home owners
plant tree buffers or rain gardens can seem like "land taking." And
when it comes to raising taxes, it can be a tough sell to get voters
to make the connection between their tap water and their lawns—
let alone the parking lot at the shopping mall. "Right through the
1970s, '80s, and '90s, rivers were seen as pipes providing water and
for treating waste," Kauffman said. "We did everything we could to
get storm water out of sight, so it could stay out of mind."

The day was getting late. As Kauffman and I pulled our canoe out
of the water, a few miles upstream from the city of Wilmington, I
noticed a pair of Baltimore orioles zipping through the trees
downstream. I tried to decide whether or not this was an omen.
Orioles are no longer a common site in the region. Was the fact
that a pair had taken up residence here a good sign? The water beneath our boats—the baseline for everything going into the orioles' systems, and our own—was not exactly pristine, but there
was a nice forest buffer, just the way Jerry Kauffman likes it.

Yet even here, by drinking water standards, the Brandywine
was already very, very dirty. A few miles downriver, and things
would get considerably less hospitable. Just south of where Jerry
Kauffman and I got out of our canoe, the Brandywine becomes,
officially, a city river, providing millions of gallons of drinking

water a day to the people of Wilmington and Chester County, Pennsylvania.

To return our boats to the rental company, we drove back through the miles of strip malls to the giant parking lot and the outfitter where we had rented out boats. The whole way back, I noticed, there was very little soil visible anywhere, let alone trees. All of these acres of pavement, one day soon, would get rained on and—like mall parking lots across the United States—shed their contaminants downhill, straight into the river. And from there, it would not be long before the water came straight through the tap.

Exploring the Brandywine upstream is one thing, Jerry Kauffman told me, but getting to know it downstream is quite another. You've seen the river sliding by the horse farms. Now you need to see it in the city's bowels.

DOWNSTREAM

The Brandywine rolls into Wilmington at a rate of about 300 million gallons a day. As it moves through the city, about 10 million gallons a day are diverted into an open-air, troughlike "raceway," then channeled into a big pipe leading directly into the Brandywine Filter Plant. The plant is old; it was designed around the time Frederick Law Olmstead designed the city's Brandywine Park, in the late nineteenth century. And like all urban treatment plants, it did what it was initially designed to do: prevent outbreaks of typhoid and cholera by removing human and animal waste from the city's drinking water. The trouble now, of course, is that typhoid and cholera have given way to far more subtle challenges.

Today, getting the Brandywine clean enough to drink is Matthew Miller's job. Miller, who serves as Wilmington's water quality manager, is the city's last line of defense. He is the man in charge of keeping everything the Brandywine has been collecting from getting into people's bodies. It's a big job.

Miller met Jerry Kauffman and me outside the gates of the Brandywine Filter Plant.

Again, it's important to understand that even before the Brandywine water enters the Wilmington treatment system, more than 13 million gallons a day have already been tainted by treated wastewater from plants in Downington, Coatesville, and West Chester, Pennsylvania. If a heavy rain causes the Coatesville plant to overflow with raw sewage, all that is here, too. So are all the raw animal wastes and all the pesticides from Pennsylvania and Delaware farms—some 80 percent of Delaware's rivers are contaminated with the agricultural pesticide atrazine alone—and all the runoff from the endless shopping strips in northern Delaware, and all the Viagra and birth control hormones and everything else that has not been absorbed by the good citizens of the region.

"We're the last people in the watershed," Miller says. "We do a pretty good job with what we have."

(As he was speaking, I remembered visiting villages in Indonesia in which the community elders live at the downstream end of a series of rice paddies. If the people controlling the upstream paddies don't release enough water, the downstream paddies fail. Since no one wants to upset the elders, the water keeps flowing. I wondered how differently things in Delaware would look if the DuPont family estates had been built downstream of the DuPont chemical plants, rather than vice versa.)

This is a water treatment plant, not a wastewater treatment plant, and even at the level of employee emotions, this makes a big difference. Wastewater is treated only for "aquatic and recreational" purposes, Miller says. "If you mess up wastewater, you kill a few fish," he says. "Here, if you mess up, you kill a few people."

So. Here you are, Matthew Miller, worrying about how you keep all the myriad contaminants that pour off our yards and roads and industries and farms from getting into our drinking water. How do you do it? You start by acknowledging that nothing, not chemistry or physics or biology, can return this soup to the pure form it took when it fell from the sky. Remember all the algae in the water, formed because of all the nitrogen-rich fertilizer draining off people's lawns? You try to get that out by dumping in powdered carbon. The algae sticks to the carbon, as do a number of toxic chemicals. (In Tennessee they use carbon to purify their whiskey, Kauffman says; in Wilmington they use it to purify their Brandy-wine.)

What about all those nasty chemicals and heavy metals, like PCBs and mercury? You try to get them out by dumping in liquid "ferric" (a solution containing iron), which binds to the toxins and settles out as a sludge. The sludge gets dumped into a sewer, goes through a wastewater treatment plant, and is then combined with fly ash left over from coal-fired power plants and dumped into a landfill. Not that it stays there; nature being nature, the sludge eventually, inevitably, seeps out of the landfill as a "leachate," and reenters the watershed. And around and around it goes.

But let's not get ahead of ourselves. Because all that ferric makes the water acidic, Miller and his crew have to add lime to get the pH back up to 6.5 or 6.6. So far, river water plus carbon plus ferric plus lime. And we're just getting started.

The inside of the Brandywine treatment plant reminded me of nothing so much as a fifteenth-century Hungarian bathhouse. In both places, you enter a door, go down some stairs, and find your-self in unimaginably huge underground caverns. At the treatment plant—which is only one century old, not six—what you find first is a cement floor, about eighty yards by twenty yards, with over-head skylights, built over an enormous underground tank. If you peer through a rectangular hole beneath your feet, you can see what Wilmington's drinking water looks like at an early stage of treatment. It looks like . . . well, it looks like crap. I mean this lit-erally. It's not, of course—the numberless clumps of brown sedi-ment suspended in the water are actually bits of coffee-colored "flocc," as the chunks of contaminants bound to the ferric are known. This is exactly what Matthew Miller, dipping in a clear plastic ladle attached to a four-foot pole, likes to see.

"This is good," he said. "This is very good."

"That's your drinking water," Jerry Kauffman said.

What was happening beneath our feet is called "flocculation."

The ferric-loaded water moves over a series of baffles, which screen out some of the flocc. The water is then channeled into an enormous cavern, divided into ten lanes like a giant indoor swimming pool. These are the sedimentation basins, where the flocculated sediment settles out as "mixed liquor." With all the big chunks out, the water is treated with chlorine, and then more lime, then sodium fluorosilicate, the fluoride added to protect your teeth. This last addition, for some reason, struck me as slightly unnerving, perhaps because the bags of fluorosilicate stacked in a storage room were marked not only with the word TOXIC but with a skull and crossbones. Granted, this is a question of quantity, not just toxicity. But still . . .

Fluoride has been added to municipal water supplies since 1945—making it one more chemical additive introduced to our bodies since World War II. More than 160 million Americans drink water that is treated with fluoride, and perhaps not surprisingly, the practice of adding fluoride to drinking water has come under intense scrutiny over the years. Though fluoride does occur naturally, it has also been used as a rat poison; hydrogen fluoride is regulated as a hazardous toxin in chemical plants. Employees at the EPA have asked Congress to take a closer look at some of the ingredients of the fluoride added to water, which often includes waste products from the fertilizer industry, especially since federal agencies are "actively advocating that each man, woman and child drink, eat and bathe in these chemicals."

In 2001, the CDC concluded that adding fluoride to water saved less than one decayed tooth per person. Some dental experts suggest that fluoride is best applied with toothpaste, rather than through the bloodstream. In Europe, where the drop in tooth decay has paralleled that in the United States, seventeen of twenty-one countries have either refused or discontinued fluoridation.

After the fluoride, zinc orthophosphate is added as a corrosion inhibitor, which is thought to create a film inside the city's water pipes that prevents calcium and magnesium from rusting, coating, or clogging the pipes. (It turns out that most of customers' complaints about the city's water are about the discoloration caused by rusting pipes.)

Then comes what Jerry Kauffman calls "the nasty stuff," liquid sodium hypochlorite. After September 11, 2001, the city upped the dosage to assuage anxieties about someone dumping toxins in the water supply. (Not that chlorine is without its problems. Miller recently fielded a call from a woman who complained that chlorine levels had gotten intolerably high. "The water is burning me," she said. But if you're Matt Miller, what are you going to do?)

After chlorination, the water is pushed outside, to open-air concrete "settling tanks," whose twenty-four square chambers reminded me, for some reason, of a catfish farm. It was at this point that the whole process of municipal water treatment finally struck me as absurd. Not only are the outdoor tanks literally crumbling—I saw rusty rebar sticking up from between decaying dividers—they are bombarded by what Miller delicately called "bird crap."

The tanks "are open to all the birds who would like to relieve themselves into them," Miller said. "That's a disadvantage. It's the same stuff we tried to take out in the first place." To fend the birds off, the city had hung a single plastic owl from a chain.

From the catfish ponds, the water is moved back inside. In a room equipped like a high school science lab, nozzles from hoses attached to each stage of the treatment process pour into Big Gulp cups from 7-Eleven. The water is tested every couple of hours for everything from turbidity (cloudiness) and pH to chlorine and fluoride, to make sure fluctuations in things like rainfall and runoff don't throw off the numbers. Visually, the difference between the "raw" river water and the final "finished stage" was striking: what typically comes in at 7.26 turbidity units has now settled out to below .3, and often as low as .005.

The water is then moved into another cavernous, beautifully tiled room. It looks like your kitchen might look, if it was equipped with an Olympic-sized pool. This is where the water goes through final chlorination and filtration, seeping through beds of sand and into a clear well. From there, it's off to the customers.

The story doesn't end here, of course. The water still has to get to your tap, often through pipes that were manufactured a long time ago. Some were made with lead solder. In the Maine body burden study, people's lead contamination came not just from old lead paint but from lead solder used on water pipes. The presence of lead in creaky municipal water systems is an especially vexing problem in older cities like Baltimore, which still relies heavily on an ancient water distribution system, from the city's main trunk lines to the pipes inside older homes. And as with other contaminants, running water through a municipal treatment plant does not remove all the lead. Neither does "purging" the water run from your tap first thing in the morning, since this sheds only the water touching the pipes closest to the tap itself, and not the water deeper in the system, where it is at least as likely to be contaminated. Besides, even flushing for fifteen minutes is only temporary, since lead levels rise again after just fifteen to thirty minutes once the tap has been turned off.

Other pipes were lined with toxic chemicals. (In the 1960s, the Johns Manville corporation installed a thousand miles of cement water pipes in New England, in cities like Providence, Rhode Island, and in the fast-growing villages on Cape Cod. The pipes came with a special feature: a plastic lining meant to improve the taste of drinking water, applied with a solvent called tetrachloroethylene, more commonly referred to as perchloroethylene (PCE), or "perc." Known even at the time as a "suspected human carcinogen," perc was still not regulated by the Safe Drinking Water Act. Studies would later reveal elevated rates of cancer mortality on the Upper Cape during the years the pipes were in place. In some EPA studies, more than a quarter of groundwater and nearly 40 percent of surface water sources had some degree of PCE contamination, and the compound still shows up in parts of New Jersey, New York, Arizona, and Massachusetts.)

So, I asked Kauffman, given the ignorance about rivers and the difficulty of keeping municipal water sources clean and the ancient infrastructure of most American cities, should people just throw in the towel and drink bottled water?

He smiled.

Since some bottlers simply repackage municipal tap water, people who buy "filtered" water in plastic bottles may simply be paying for what they can get for free. Regulations for bottled water don't require disinfection for pathogens like cryptosporidium or giardia. (In New York City, city officials recently spent $700,000 to convince residents that their water, which is piped in from the Catskills and considered some of the best drinking water in the country, is worth drinking.)

And given what we know about the chemicals that leach out of plastic bottles—phthalates if the bottle is crinkly, bisphenol A if the bottle is stiff—drinking from a bottle is hardly a comfort. To say nothing of the added environmental cost, in both the oil-based manufacture and the imperfect disposal, of the plastic bottles left over once the water has been consumed. By some estimates, 1.5 million tons of petrochemicals are used to make water bottles. Every year.

"Sales of bottled water are driven by anxiety," he said. "But bottled water is even less regulated than tap water. There is a list of ingredients on all other processed foods and drinks, but there are no ingredients listed on bottled water. In my opinion, if this stuff is sold as a consumer product, it should be monitored like any other food product."

In Jerry Kauffman's office, the staff follows a policy called "drinking the water"—kind of like "walking the walk." The office used to have bottled water delivered, but Kauffman canceled the contract. Some people were upset by this, but Kauffman decided that drinking what came out of the tap would keep his team focused on what goes into the tap.

For the home owner, Kauffman recommends taking simple steps, like using a good end-of-tap filter. Filters will help remove some of the metallic odors and tastes from the water, along with some synthetic chemicals that may have slipped through municipal water treatment.

But Kauffman is also quick and consistent with his recommendation that people become more aware of the way their behaviors affect their watersheds. Individual home owners, for example, can use fewer pesticides on their lawns or, better yet, replace their lawns with native trees and plants. Farmers can plant trees along stream banks to keep agricultural waste from flowing into creeks. Good landscaping and farming practices, Kauffman said, can lower the amount of nitrogen in streams from 500 parts per million to 1 part per million, and at very low cost. He recommends the proper disposal of pharmaceutical drugs, and much more awareness about the harm pavement does to river systems. Cities can be more vigilant as well. The city of Wilmington, he noted, is working to cover over the open-air tanks at its water treatment plant; has constructed huge storage tanks under the city parks to store sewage overflows during storms; and is paying Amish farmers upstream to fence their cows away from streams and reforest the floodplains.

Everything we do is connected, one way or another, to water, and ultimately our bodies will be only as healthy as our drinking water supply. "In our field, we believe people will do more to protect their water if they identify more with their drinking water sources, if they are more connected to the rivers and bays that surround them," Kauffman told me. "You know what Ben Franklin said: 'An ounce of prevention is worth a pound of cure.'"

FIVE The Lawn

For the last dozen years or so, I have taught a course at the University of Delaware called The Literature of the Land. It's one of my favorite classes, not least because of the writing requirements. In addition to asking students to read a half dozen books about a particular subject (nature and religion, industrial food production, climate change), I have them spend at least an hour every week sitting in the woods and writing about what they see. Not what they think about. What they *see*. Learning field observation, I tell them, is as important for aspiring writers as it is for aspiring naturalists. Both demand attention to detail, both demand the ability to shut up and listen. Students are required to carry a field guide with them, and to take detailed notes on the things they see springing to life or dying back.

Over the years, a pattern has emerged. At first, all the students see are "trees." A week or two later, they are seeing "white oaks" and "tulip poplars" and "sugar maples." A few more weeks, and their eyes are more fully accustomed to the pace of life in the woods. They move through the space more slowly, and more quietly. They see foxes. And great blue herons. And barred owls.

My goals, not usually articulated until well into the semester, go beyond writing instruction, of course. What I'm trying to do is encourage students to spend some time, every week, turning down the noise in their lives. I want them to try—just for a time, but every week—to unplug from their cell phones and laptops, to disentangle themselves from the social and academic pressures of college, to take a walk in the woods. This also offers, not incidentally, an opportunity for self-reflection, for meditation, and for a renewed sense of engagement with the natural world.

Over the course of a dozen years teaching this class, something has become clear to me. Every semester, more and more students enter my classroom having spent virtually no time outdoors. They have not camped or climbed trees. They have not backpacked. They have not paddled rivers. They don't fish, or hunt, or climb mountains. Not with their families, not with their friends.

A couple of years ago, on the first day of class, I asked my students what words they think of when they hear the word "wilderness." In past years, they offered both concrete, experiential words ("my family's trip to Yosemite") and what were surely abstractions ("jungle," "rain forest," "wild animals"). But this year was different, and for me, it marked what I still think of as a benchmark moment. What words did this group of students have me list on the board? What did wilderness mean to them?

"Rape," said one.

"Fear," said another.

"Loneliness," said a third.

What was going on here? I knew most of these kids were suburbanites, but hadn't they ever been Girl Scouts? Or gone to a summer camp? Where were they getting their news about the natural world? Nature, for these students, had somehow become an abstraction at best, and a source of anxiety at worst. There was no sense of joy, or adventure, or mystery. No wonder. Forget about fantasies of strapping on a backpack after college and hiking the Appalachian Trail, paddling the Mississippi River, or bumming around Alaska. These kids seemed afraid to leave their backyards.

I mention this because it has begun to seem that the degree to which we have become abstracted from our natural surroundings might tell us a lot about the degree to which we are willing— consciously or not—to live with synthetic chemicals. If a college student can't tell you what river or reservoir his tap water comes from, why would he ever stop to think about all the toxins that water might be collecting on its way to his tap? If she has no concept where the meat in her Big Mac came from, what possible reason would she have to consider buying food grown on a local farm, or on land that has not been sprayed with pesticides? If his concept of a beautiful college campus is unbroken stretches of perfectly groomed lawns, why would he possibly object to the acres of toxic herbicides that such lawns require to look so pristine? Likewise, this same student has likely never considered that these chemicals will wash off the lawn at the first rain, will flow into a storm sewer, and will end up, eventually, somewhere in a river, where they will at least damage aquatic life and will probably make their way back into his own (or someone else's) drinking water. But why would he think about this? He can't see the creeks, because they are all buried underneath what he thinks of as a Frisbee field.

One recent spring, fifteen students and I spent most of the semester talking about environmental contaminants in our food, in atmospheric greenhouse gases, and, via toxic chemicals, in our everyday lives. On a beautiful April day, we decided to meet outside, on the campus's central lawn, right between two classroom buildings that just happened to bear the names of two gargantuan chemical companies: the DuPont building was to our left, the Gore (as in the Gore-Tex chemicals and textiles company) building to our right. In the middle of a conversation about agricultural pesticides, a university groundskeeper, dressed from his feet to his neck in a white chemical suit, drove by on his rider mower. He wasn't cutting the grass, though; he was spraying it. And not from one nozzle, but from a half dozen. Up and back he went, describing parallel lines as neat as those in any Iowa farmer's cornfield. Up and back. Up and back. Not a blade of grass escaped the spray.

This became what is known as a perfect teaching moment.

Who's going to go up and ask him what he's spraying? I asked.

My students smiled nervously at one another, until a bright young woman volunteered. We watched as she marched over to the groundskeeper and waved cordially until he turned off his engine. They spoke for a minute or two, out of earshot, and she returned.

"He said he's spraying 2,4-D," she said, and it was clear that she had no idea what this chemical was. "He said we didn't need to worry, because he sprayed the place where we're sitting at five A.M. this morning."

About seven hours earlier. My students chuckled self-consciously. Seven hours? He's wearing a full-body chem suit, and we're sitting on the grass, seven hours later, in shorts and bare feet?

My students had never heard of 2,4-D, otherwise known as 2,4-dichlorophenoxyacetic acid. But they had heard of Agent Orange, the notorious defoliant used in Vietnam. This pesticide, 2,4-D, I told them, is a constituent of Agent Orange.

My guess is that except for my students, who happened to be studying pesticides, few of the thousands of people walking across the great expanse of lawn that morning noticed what was happening, let alone cared. It was a beautiful day, the sun was shining, the grass was green. What's not to like? The guy in the white suit, on a machine with nozzles going full tilt, is as much a part of the campus landscape as the brick walkways and the grass itself.

Back in our little circle on the grass, I told my students what I knew about 2,4-D. I told them, for one thing, that it is the most extensively used herbicide in the history of the world. I also told them that it was developed secretly during World War II, mostly as a weapon to destroy an enemy's rice crops.

Yet despite its military history, 2,4-D has long been presented by the lawn-care industry as safe for civilian use. Some of the evidence, in retrospect, seems less than convincing.

In the 1940s, E. J. Kraus, the head of the botany department at the University of Chicago, said he fed five and a half grams of pure 2,4-D to a cow every day for three months. The cow was fine, he said, and so was her calf. Kraus said he himself had eaten a half gram of the stuff every day for three weeks, and felt great. This was apparently good enough science for the rest of the country; within five years, American companies were producing 14 million pounds of 2,4-D a year. By 1964, the number had jumped to 53 million pounds. "The public's enthusiasm for the newly unveiled herbicide was equaled by universal ignorance of the chemical and how it worked," a 1967 history of the compound reported. "Even without clearly understanding how 2,4-D worked, it looked good to 'weed men.' "

Here's why: beyond its ability to defoliate jungles, 2,4-D proved to have another property that had commercial lawn companies seeing green: it could kill broad-leaved plants, like dandelions and clover, without killing grass. Today, annual sales of 2,4-D have surpassed $300 million worldwide. Since it does not require a license to buy or to use, 2,4-D can be found in many "weed and feed" products, like Scotts Green Sweep, Ortho Weed B Gon, Salvo, Weedone, and Spectracide.

At first, 2,4-D's impact on human health seemed fairly tame— skin and eye irritation, nausea and vomiting, dizziness, stiffness in the arms and legs—and conventional lawn-care companies have long dismissed health worries. The amount of chemicals in most lawn sprays are so diluted, the companies say, that "even if it were the most toxic material known to man, in the solution we use it, it would be harmless."

But as with most chemicals, the effects of 2,4-D are more worrisome when they are considered over time. Because it is designed to mimic a plant's natural growth hormone, it causes such rapid cell growth that a plant's normal transport systems become destroyed by abnormally fast tissue growth. The stems of plants treated with 2,4-D tend to become grotesquely twisted; the roots become swollen; the leaves turn yellow and die. Plants quickly starve to death.

Given its effects on cell growth in plants, it should perhaps not be surprising that 2,4-D has also been shown to disrupt human hormones. The National Institute of Health Sciences lists 2,4-D as a suspected endocrine disruptor, and several studies point to its possible contribution to genetic mutations and problems with re-productive health. Although the EPA continues to maintain that there is not enough evidence to classify 2,4-D as a carcinogen, a growing body of research has begun to link it to a variety of cancers, including non-Hodgkin's lymphoma. A 1986 National Cancer Institute study found that farmers in Kansas exposed to 2,4-D for twenty or more days a year had a sixfold higher risk of developing non-Hodgkin's lymphoma. Five years later, another National Cancer Institute study showed that dogs were twice as likely to contract lymphoma if their owners used 2,4-D on their lawns. Like flame retardants and countless other compounds, 2,4-D also tends to accumulate *inside* people's homes, even days after the lawn outside has been sprayed. One study found 2,4-D present in the indoor dust of 63 percent of sampled homes; an-other showed that levels of 2,4-D in indoor air and on indoor sur-faces like floors and tables increased after lawn applications. Exposure levels for children were ten times higher than before the lawns were treated—an indication, among other things, of just how easily the chemical is tracked inside on the little feet of dogs, cats, and children.

Of course, 2,4-D is just one of scores of pesticides in broad use today. Dr. David Pimentel, a professor of entomology at Cornell, has written that 110,000 people suffer from adverse health effects from pesticides each year, and that 10,000 cases of cancer may be attributable to pesticide exposure.

All of which makes running around in bare feet—let alone sit-ting outside for spring seminars—a little less appealing. In the weeks that followed our encounter with the spray nozzles, I en-couraged my students to think a little harder about toxic chemi-cals while completing their journal assignments.

The student who initially spoke to the groundskeeper decided to look into the EPA's toxic release inventory for her home state of Massachusetts. She discovered that in 2006, 4.8 million pounds of toxic waste were deposited in-state, and an additional 2.1 million pounds were transported out of state for disposal.

Another senior wrote that her newfound awareness of toxins

in the environment had left her craving ignorance, if only for a day or two. Her mother, a college professor, is a four-time breast cancer survivor. Another wrote that his father had been treated for prostate cancer, as had one of his uncles. Another uncle had testicular cancer. His grandparents died of lung cancer. His mother has had lymphoma removed from her eyes, and her thyroid from her throat. "That's how commonplace cancer is now," the student wrote. "That's what we've been reduced to: future patients."

A pretty disheartening way to view what once looked like a nice place to play Frisbee.

So how did we come to this?

A hundred years ago, 60 percent of Americans lived in rural country. Today, 83 percent live in cities or suburbs. Along with all that new housing has come an astonishing shift in the landscape. Over the last half century, Americans have become obsessed with grass. When you add up the country's 58 million home lawns and 16,000 golf courses, you get close to 50 million acres of turf in the United States. That's a national lawn roughly the size of Nebraska. And that number is growing by 600 square miles a year. Think about that the next time you are worrying about the disappearance of the Amazon rain forest.

I'm not sure if 50 million acres sounds like a lot or a little. But when you consider the total amount of herbicides and pesticides it would take to chemically "manage" a lawn the size of Nebraska, you get a sense of what we're talking about. Americans spend some $40 billion a year on lawn care, more than the gross national product of Tunisia. Estimates of the number of American households that use pesticides run as high as 82 percent. And this doesn't even take into account the fuel required to mow Nebraska. On average, mowing your lawn for one hour produces as much pollution as driving your car about 650 miles; in an average year, Americans burn 800 million gallons of gasoline in their lawn mowers. And each time we fill the gas tanks on our lawn mowers, we spill a few drops on the lawn or the driveway. If you tally up all the little spills made by all the people mowing their lawns in the United States, you get a cumulative gasoline spill of some 17 million gallons a year—roughly 50 percent more than was spilled by the *Exxon Valdez*. And blowing the leaves? Using a gas-powered leaf blower for a half hour releases as much carbon as driving a car 7,700 miles at 30 miles per hour.

In other words, a national landscape that Americans still imagine to be a great wilderness has, in a few short decades, turned into an enormous source of synthetic chemicals.

And like most synthetic things, the suburban American lawn has its roots in the postwar 1940s. With GIs returning from the war by the hundreds of thousands, developers like Abe, Bill, and Alfred Levitt reacted to the explosion in demand by building homes as if an assembly line. Within a couple of years after the war, Levittown, Long Island, went from a potato field to a full-blown suburb with more than 17,500 homes—each on one-seventh of an acre. Only 12 percent of each Levittown lot was taken up by a house. The rest was devoted to landscaping. There were a few fruit trees and a couple of evergreens, but everything else—times 17,500—was grass. "A fine lawn makes a frame for a dwelling," Abe Levitt said in 1949. "It is the first thing a visitor sees. And first impressions are lasting ones."

Especially for a country so unsettled by a world war, the consistency and conformity of the Levittown vision felt comforting. As the fear of fascism gave way to a fear of communism, the state of one's lawn was considered a public declaration of one's allegiance to the American Dream. In a 1948 newspaper column, Abe Levitt wrote that a manicured lawn was the clearest possible sign that a home owner was both honorable and trustworthy—that is to say, not a Communist. "Remember, your lawn is your outdoor living room about 7 months of the year," he wrote. "Your visiting friends form their opinions of the neatness and cleanliness of your house at their first approach." Neighborhood covenants required home owners to mow their lawns at least once a week between April and November.

The May 1957 cover of the *Saturday Evening Post* showed four families reposing on identical postage-stamp squares of grass, separated by identical lengths of white picket fence behind identical tract houses. All of the people were white (until 1960, Levittown's 80,000 residents included no African-Americans), and they were all engaged in some form of "gardening"—a man pushing a lawn mower, a woman walking, apparently barefoot, across grass that looks as manicured as the felt on a billiard table. This was the postwar suburban dream: lawns became as controlled and as sanitized as the interiors of the homes themselves. In suburbia, *Time* magazine noted in 1959, crabgrass on a lawn can lower a man's status "faster than a garbage can in his foyer." The weed "has become a neighborhood problem like juvenile delinquency. If not snuffed out in one spot, it quickly spreads to another."

Connecting (and encouraging) all these new suburbs, of course, was the country's brand-new interstate highway system, and along with the four-lane highways came 200 feet of right-of-way, or 25 acres of turf per mile. By 1961, this added up to the equivalent of some 29 million football fields—all of which needed regular mowing and regular spraying. In 1962, Rachel Carson noted that an area larger than New England—50 million acres— was under management by utility companies and that much of that land was treated chemically for "brush control." In the Southwest, an additional 75 million acres were managed with chemical sprays. "Chemical weed killers are a bright new toy," Carson wrote. "They work in a spectacular way; they give a giddy sense of power over nature to those who wield them, and as for the long-range and less obvious effects—these are easily brushed aside as the baseless imaginings of pessimists. The 'agricultural engineers' speak blithely of 'chemical plowing' in a world that is urged to beat its plowshares into spray guns."

The trouble with a grass obsession is that grass, strictly speaking, is an early stage of ecological development—especially since the grasses we have planted here are not even native to North America. Kentucky bluegrass comes from Europe and northern Asia; Bermuda grass comes from Africa; zoysia grass, from East Asia. What this means is that lawns actually want to be invaded, by everything from weeds to wildflowers to trees. Since nature trends toward diversity, not uniformity, keeping a patch of land exclusively in grass requires ceaseless human intervention, as any suburban teenager knows: lawns have to be mowed and sprayed and reseeded constantly. It takes an enormous amount of energy to keep a lawn the size of Nebraska in good health, especially on a continent where lawn grass itself is an exotic guest.

Keeping a patch of grass from doing what it wants to do—turn into a field, and then into a forest—requires a lot of work. But Americans have been up to the task: in just a dozen years, from 1946 to 1959, sales of lawn mowers went from 139,000 to 4.2 million. By the 1960s, Americans were so worn out from all the time they devoted to keeping their lawns intact that they did what anyone does when they get exhausted: they paid someone else to do the work. And a lot of work there was.

In 1968, a former garden store owner and sod farmer in Troy, Ohio, named Richard Duke got so tired of fielding questions about lawn care that he decided to start his own company, which he named ChemLawn. His signature technique: rather than luring customers into a store to buy lawn chemicals, he hired "specialists" to drive to the customers' homes and spray liquid chemicals—a mix of synthetic fertilizer, pesticides, and herbicides—from their trucks. "The ChemLawn truck, a custom-designed six-wheel tanker with a huge lawn-green logo splashed across its sides, would come and go while residents were at work and at school," a profile of the company reported. "The only sign of its passing was a notice on the door reminding people not to roll around in the grass until the chemicals had had a chance to dry (about an hour)."

What ChemLawn seemed to be selling was more than a kind of exterior decorating; it was selling consistency, even comfort. The lawn "specialists" were like doctors who made house calls, promising to rid lawns of pests and disease.

"Saturday was a big sale day," said Richard Lyons, who rose from a sprayer to vice president. "More than once, we'd park in a cul-de-sac, and people would come to line up at the truck and sign up for the service. You'd see them walking from blocks away." The business model was a smashing success: started with $40,000 to outfit a couple of spray trucks, the company was making $300 million a year by 1985.

By 1999, more than two-thirds of America's home lawns were being treated with chemical fertilizer or pesticides—14 million of them through a professional lawn-care company. A year later, the federal General Accounting Office reported that Americans were spraying 67 million pounds of synthetic chemicals on their lawns every year, and that annual sales of lawn-care pesticides had grown to $700 million. Lawn-care companies were doing an additional $1.5 billion in business.

All those trucks rolling around our suburban neighborhoods seem to represent something more than a communal desire for soft carpets of monoculture grass. They seem to represent a relief from anxiety. (Why else call a company "Lawn Doctor"?) But anxiety from what, exactly? Perhaps we haven't moved so far from Levittown, after all. Hiring trucks full of lawn-care "specialists" is, for one thing, an unusually public declaration that you have the money not to take care of your yard yourself.

Property values are clearly associated with high-input green-lawn maintenance and chemical use, researchers from Ohio State reported in 2003. Strangely, the researchers found, wealthy home owners continue to use lawn chemicals even when they are conscious of their harmful effects—even those who claim to be concerned about community, family, and the environment. "Lawn chemical users typically associated moral character and social reliability with the condition of the lawn, suggesting that the lawn represents a *public* statement about proper *private* behavior," the study reported. In other words, a chemically treated lawn still means a house is not hiding Communists.

Lawn-care companies have spent so many years marketing their chemicals that tending lawns without them has come to seem quixotic at best. Type "organic gardening" into the Google search engine, and the first thing that pops up is a link to Scotts Miracle-

Gro, perhaps the most famous synthetic chemical fertilizer ever put on the market.

But don't expect Scotts to sit and wait for you to find them. The company sends our house a mailing every spring, urging us not to fall behind our neighbors. "What's wrong with your lawn?" the flyer asks. "What do you need to do now to protect your lawn from unsightly weeds, insects you can't even see, and damaging turfgrass disease? Call Scotts LawnService like so many of your neighbors who had those problems. You'll see us treating their lawns throughout the season."

This is designed, of course, to cause you anxiety, and then, if you use their products, to offer you comfort. Included in the mailer are before and after photos, showing a lawn that miraculously changes from a shabby patchwork of green and brown to a green mat as uniform as Centre Court at Wimbledon or the eighteenth green at Augusta. There are also plenty of customer testimonials. "Our lawn is looking better and better," says one Robin M. "We get so many compliments on it and we want the lawn to continue to look its best. A lawn has many challenges and Scotts has been a real plus in our lawn's overall health and appearance!"

So: Is your neighbor's lawn greener than yours? Less weedy? What might this mean? Is their net worth higher than yours? Are their kids smarter? Is their sex life better? A TruGreen flyer we got this year features a photo of a smiling young man sharing a patio moment with not one, not two, but three beautiful women. If that comes with the lawn treatment, who could refuse?

Like countless other lawn-care companies, Scotts promises visits to your home by its team of "professionals" who are "trained" by Scotts LawnService Training Institute. The "professionals" will analyze and diagnose your lawn's "problems" and then recommend a course of chemical therapy. "No common, agricultural fertilizers are ever used," the flyer notes. "We apply Scotts professional slow-release fertilizer—the same ones [sic] developed, tested, and used on tens of thousands of lawns." A basic regimen might include regular applications of Scotts® slow-release fertilizers, plus Scotts® Halts® Crabgrass Preventer, Ortho® Weed B Gon® Pro™

Dandelion & Weed Control, and Ortho Max® Pro™ Guaranteed Season-Long Insect Control. A Plus package gets you all that plus Scotts® GruEx® Guaranteed Season Long GrubControl and some "core aeration," which means a guy pushes a roller around your yard, punching holes in it. The Complete package gets you all this plus some lime for pH balancing and some extra grass seed.

Even grass seed, nowadays, comes coated with chemicals. A bag of Scotts grass seed is labeled "Pure Premium," as if it's a cut of good beef. A closer look reveals that the seed has been treated with Apron XL LS fungicide, whose scientific name is, and I quote, (R)-Z-([2,6-dimethylphenyl]-methoxyacetylamino)-propionic acid methyl ester. Perhaps because of the double parenthetical, the bag helpfully notes that the fungicide is "commonly known as Mefenoxam or Metalaxyl-M." That does help. The bag also requests, in bold black type, that it be kept out of the reach of children, and that anyone using the seeds "wear long-sleeved shirt, long pants, shoes, socks and waterproof gloves," and that the gardener "remove contaminated clothing and wash before reuse." The bag also provides a phone number for "emergencies" related to the fungicide: call Syngenta Crop Protection, 1-800-888-8372.

Did you know that grass seed could contaminate your clothes?

Anxiety about the toxicity of lawn chemicals has presented companies with a significant public-relations challenge. In Baltimore, Scotts ends a radio commercial with the request that home owners "remember to sweep fertilizer off your driveway to keep drinking water clean." This gesture seems disingenuous at best; sweeping chemicals from a driveway back onto a lawn would prevent a fraction of the treatment from washing away during heavy rains. And as Jerry Kauffman would be quick to point out, all of this stuff ends up, one way or another, in our water supply.

As the use of chemicals has become more and more widespread, chemical companies have realized an unexpected source of profits. Herbicides like 2,4-D preserve grass but kill clover; that's why people buy them. But clover, unlike grass, can pull nitrogen out of the air and "fix" it in the soil. Without clover, soil becomes nitrogen-poor. So what did the chemical companies do?

They offered to replace the depleted nitrogen, which home owners used to get for free, with synthetic nitrogen, for which they
have to pay.

And pay. In the Chesapeake Bay watershed, as in watersheds
all over the United States, nitrogen runoff is considered among the
worst problems for water quality. There are 3 million acres of turf
grass in the Chesapeake watershed, most of it fertilized with synthetic fertilizer. Since synthetic fertilizers are water-soluble, a good
bit of it will run off your lawn after a rain—which is a waste of
your money and creates genuine problems in river water once
your fertilizer mixes with the runoff fertilizer of thousands of
other homes. This is because fertilizer doesn't just feed the grass
on your lawn; it also feeds plants that grow in the water. Doused
with chemical fertilizers, underwater algae can grow exponentially, creating "algae blooms" that—as they die and decay—suck
most of the oxygen out of a lake or a bay. Some estimates say that
each pound of phosphorous (which also comes from things like
conventional laundry detergents) entering the water produces
twenty pounds of algae; the cost of removing twenty pounds of
algae is roughly $200. Now start multiplying.

River-protection groups talk about pollution "loads" that build
up from the rain running off farms, roads, and households. The
thousands of pounds of fertilizing chemicals then create massive
"dead zones," in which nothing—not fish, not plants, nothing—can
grow. This is as true in the Gulf of Mexico as it is in the
Chesapeake. In 2007, the Chesapeake Bay Foundation offered its
annual report card on the bay's health, and the intertwined threats
show just how much trouble chemicals can pose. The bay got an F
for nitrogen pollution, a D minus for phosphorous, an F for water
quality, an F for dissolved oxygen, and a D for toxics. All told, on a
scale of 100, the bay's health was rated at a 28.

In California, scientists are discovering something even more
haunting. Algae blooms off the coast are not only removing oxygen from the water, they are releasing a toxin that appears to be
causing sea lions to go into epileptic seizures. The toxin, domoic
acid, enters the food chain when sardines and herring eat algae,

then moves into the amniotic fluid of sea lions who eat the fish. If the sea lion happens to be pregnant, her fetus can be contaminated by the toxin and, years later, develop epileptic seizures. Documentary footage of a sea lion in seizure is something you need to see only once.

And it's not like Americans enjoy manicured lawns only at home. They also create them in order to play that most chemical-dependent of all pastimes: golf. By 2004, there were just under 15,000 golf courses in United States—all told, a patchwork of chemically treated turf the size of Rhode Island and Delaware combined. A 1990 survey of 52 golf courses on Long Island, conducted by the New York attorney general's office, found the courses covered with 50,000 pounds of active ingredients—about 18 pounds per acre per year, or six times what was needed to control weeds. "If you scraped a golf green and tested it, you'd have to cart it away to a hazardous waste facility," said Joseph Okoniewski, a biologist with New York's Environmental Department of Conservation.

A former colleague of mine, an avid golfer and a cigar aficionado, once told me that he would never rest his cigar on the turf when he set up to hit the ball. He said he knew a man who claimed to have gotten mouth cancer from pesticides on the golf course. What I found striking about this story was that this man, a lifelong journalist and professor, was more worried about pesticides (about which the science is considered "inconclusive") than he was about smoking (about which the science is anything but).

Which brings me, inevitably, to the story of the Irish golfer who (how do I say this delicately?) liked to lick his balls. In 1997, doctors in Ireland reported the strange case of a sixty-five-year-old retired consultant engineer from Dublin who went to see doctors complaining of chronic lethargy, ink-dark urine, and acute abdominal pain. The doctors diagnosed hepatitis. After ruling out such other causes as drug or alcohol "indiscretions," the doctors came to a startling discovery. The man said he was a widower who played golf every day, and had developed a peculiar (if apparently not uncommon) habit: whenever he lined up to putt, he would clean his balls with his tongue. This despite signs on the course expressly warning against "licking balls," because of recent applications of 2,4-D. Once the doctors advised against licking, the report said, the man "ceased his habit, and within two months his liver function tests had returned to normal and he felt well."

Four months later, however, the patient returned for more liver tests. He was sick again. He told his doctors that he had grown skeptical of their diagnosis and, just to prove them wrong, had "resumed licking his golf ball." Once again, his tests showed hepatitis. Convinced at last, the man accepted a diagnosis of "golf liver," and now, the researchers reported, he plays golf regularly, carries a damp cloth to clean his golf ball, and "remains well."

Paul Tukey knows a thing or two about pesticides; the man who invented 2,4-D was a distant cousin. When Tukey was a kid in the late 1960s, his grandfather would hire a biplane a couple of times a year to spray his 300 acres of fields in Bradford, Maine. The fields were mostly planted with feed for a herd of 260 cows, not with crops intended for human consumption. The presumption, apparently, was that pesticides sprayed on the feed crops would not enter the human food chain—despite the fact that the cows were milk cows and Paul drank their milk. But none of this was on Paul Tukey's mind as he and his grandfather drove out to watch the airplanes. Spraying day was, for a young boy on a farm, a thrill.

"My grandfather would go out in the field, dressed in his wool underwear and thick heavy pants, and wave the biplane over his field," Tukey recalled. "They'd drop this white powder, and he'd get back in the truck looking like Frosty the Snowman. Then we'd drive to the next field, and he'd do it again. My grandfather was getting doused twenty times a day, but he would never let me out of the truck. I would *beg* him. I always wondered why I couldn't go out and get dusted."

Tukey's grandfather died of a brain tumor at the age of sixty.

Now tall and athletic, with sun-bleached hair, Tukey combines the easy charisma of a ski instructor with the ardor of an evangelist. And like any convert, his road to enlightenment was not exactly straight and narrow. Tukey followed the family's agricultural tradition but charted his own way through central Maine's growing suburbs. For years, he operated one of the region's largest landscaping services, tending the lawns of more than eight hundred customers. He considered his job ideal. He worked sixteen hours a day, but he worked outside, in shorts and sandals. He never bothered with protective gear.

Then one day in 1993, he started getting nosebleeds. His vision became blurry. But that spring, with business booming and a new baby boy, Tukey was too busy to worry about his health.

As his business grew, Tukey took a job tending the grounds of a hospital in Portland. He had hired a group of graduate students for the work, and one day, the students' professor, an eminent horticulturist named Rick Churchill, came by to say hello. In Maine's gardening circles, Churchill was something of a hero, and Tukey, like many of Churchill's acolytes, ached for the elder man's approval. Tukey reached out to greet him.

Churchill's eyes remained focused on the ground. The weeds, which Tukey and his team had recently doused with synthetic herbicides, had begun to curl up and turn brown. Churchill gave Tukey a dismissive grunt and turned his back. He wouldn't even shake Tukey's hand.

"I didn't mince words," Churchill said. "I asked him how anyone in good conscience could be applying pesticides on the grounds of a hospital where there were patients being treated for cancers that could be linked to their exposure to pesticides. I asked whether he knew anything about the toxicity ratings of what he was applying, and how dangerous many of these compounds were to an individual compromised by illness. When I left his demeanor had changed."

Churchill's words cut deeply. "It was devastating psychologically," Tukey told me. "In Maine, Rick Churchill is an icon. His approval was a stamp that we all craved. It was especially daunting for me way back then because I had no idea at all how to get rid of those weeds without the chemical weed killers."

Tukey was stuck. Here he was, with twenty-eight employees, a growing family to feed, and a business that relied fundamentally on toxic chemicals. What was he supposed to do?

"Right away that set up a good-versus-evil dynamic that is still inherent today in the industry," Tukey said. "By that, I mean that landscapers like me had to do bad things to the environment or health to get the good result that our customers demanded. It's not easy. Landscapers are stuck with a quandary: put down poisons or displease the customer."

Tukey started doing some research. What he found was troubling: Pediatric cancers in Los Angeles had been linked to parental exposure to pesticides during pregnancy. In Denver, kids whose yards were treated with pesticides were found to be four times more likely to have soft-tissue cancers than kids whose yards were not. Elsewhere, links had been found between brain tumors in children and the use of pest strips, lindane-containing lice shampoos, flea collars, and weed killers.

Tukey also learned that exposure to lawn chemicals was particularly alarming for people—like him—who sprayed lawn chemicals for a living. One study showed a threefold increase in lung cancer among lawn-care workers who used 2,4-D; another found a higher rate of birth defects among the children of chemical appliers. Tukey, who had been literally wading in synthetic chemicals for nearly two decades, had never given much thought to the impact these chemicals might be having on his own health, let alone his family's. When his chronic skin rashes and deteriorating eyesight finally landed him in a doctor's office, he learned that he had developed multiple chemical sensitivity. Not only this, but his son—who had been conceived in 1992, during the height of Tukey's use of synthetic chemicals—had just been diagnosed with one of the worst cases of ADHD his doctor had ever seen. Tukey remembers this diagnosis as "a sledgehammer in the gut."

"All the evidence indicates that you don't want pregnant women around these products, but I was walking into the house every single night with my legs coated with pesticides from the knees down," he said. "Even when my son was a year or two old and would greet me at the door at night by grabbing me around the legs. He was getting pesticides on his hands and probably his face, too."

And then it happened, the evangelical moment.

Tukey was driving around in his truck when he noticed a sign: a local department store was having a two-for-one sale on Scotts Turf Builder. Tukey made a beeline to the store. He was going to buy the store's entire stock. Once inside, he had little trouble finding what he was looking for; as he has said, a blind man can find the lawn chemical section in any store. The stuff is volatilizing right there in the aisle.

Walking toward the smell, Tukey noticed a young woman standing by a display of lawn chemicals. At her feet, a little girl was making sand castles from what had spilled from a broken bag of pesticides. Suddenly, in Tukey's head, everything changed. The DDT squirting from planes over his grandfather's fields, the chemicals he'd been spraying outside the hospital, and now a child playing in a pile of pesticides. Something in Tukey burst.

"I said, 'Ma'am, you really shouldn't let your child play with that stuff. It's not safe,'" Tukey told me. "I'm fundamentally shy, but this just came out of me."

The woman looked at Tukey, sizing him up. Here was this tall, grass-stained man, with several days of beard sprouting from his chin, glowering at her with a fire that seemed somehow too intense for someone hanging around the lawn-care section. The woman looked Tukey in the eye.

The store wouldn't sell the stuff if it wasn't safe, she said. She gathered up her child and walked away.

A store manager came up to Tukey and asked him if there was a problem. Tukey said indeed there was.

"*You* have the problem," he told the manager. "You have broken bags of poison on the floor. All those bags say, 'Keep out of reach of children'!"

The manager looked at Tukey dismissively. Those bags have those labels because of some government formality, the man said. The stuff in there isn't really dangerous. Our store wouldn't carry something that wasn't safe.

"That really was the stake in the heart of my chemical career," Tukey said. "By then, I had already made myself sick. I had already been questioned by Rick Churchill, and I had the gnawing guilt of having my employees applying the same pesticides that had made me sick. When I saw that little girl making sand castles out of the pesticides was just a sudden gut-level reaction that I couldn't have anticipated. I'm a mellow guy, but I found myself getting angry at the store manager because of his ignorance. I was shaking when I left the store. Literally shaking."

Tukey drove back to his company and issued a decree. It was time to start weaning the company—and the customers—off synthetic chemicals. The company was going organic. He wrote to his customers. Most of them seemed fine with his decision, with two provisions: You can do whatever you want, they told Tukey. Just as long as it doesn't cost any more, and as long as our lawns continue to look the same.

"What gets me, though, is this," Tukey said. "Many of the same people who gasp audibly when they hear the story of the girl playing with the pesticides actually use those same exact products on their own yards, or condone the use of those products around schools."

Quitting an addiction to toxic chemicals, in other words, was not as simple as giving them up himself, Tukey found. He also had to convince his customers, who had been saturated by chemical lawn-care advertising for decades. True, in recent years chemical lawn care had been held under the microscope, not least for marketing products in a way that seemed intended to dismiss any worries about toxicity. As early as the 1980s, New York's attorney general's office had sued the ChemLawn company for false advertising; at issue were ads claiming a child "would have to swallow the amount of pesticide found in almost 10 cups of treated lawn clippings to equal the toxicity of one baby aspirin." A few years later, the same office forced Chevron, which makes Ortho products, to stop using television ads showing barefoot children playing on lawns being treated with pesticides—all while a voice-over said, "Sure, I care about this yard, but I care about my family using it, too." Chevron paid a $50,000 fine, but admitted no wrongdoing.

In 2003, TruGreen ChemLawn tried a new approach: the company mailed flyers to children around the country, offering financial support for youth soccer programs if their parents signed up for lawn-care service. Whatever business the promotion drummed up, it also generated significant controversy. "It's sick," said a spokesman for the Toxics Action Center, a Boston environmental group. "They market through your kids, and by spraying your lawn with this stuff you are putting your kids at risk."

Even Long Island, the site of the original Levittown, has begun confronting its lawn legacy. One day in the mid-1990s, a woman named Nicole Hudson was working in her backyard while her baby was upstairs in a crib with the window open. She looked over and saw a company spraying in the next yard, then heard her baby crying. When she went upstairs, she found a pesticide mist in the room. Five days later, when she had the room tested, residue was still present on the windowsill, so she sued the contractor in small claims court and won several hundred dollars.

"The lawns out here may look good, but they are unhealthy," said Neal M. Lewis, the director of the Long Island Neighborhood Network, an environmental group. "It's steroid turf, all pumped up with drugs. People are growing chemically dependent lawns, and the chemicals are by definition poisons, designed to kill." After nine years of lobbying, a local protest gathered so much steam that New York became the first state in the country to pass Neighbor Notification legislation. Twenty other states quickly followed suit. Suddenly, it seemed, the lawn's "family value" was beginning to trade a bit lower.

To Paul Tukey, this new ambivalence about toxic lawn chemicals presented an opportunity. In 2007, he founded SafeLawns.org, a coalition of both non- and for-profit organizations that has become one of the country's leading advocates of environmentally responsible lawn care. The group's mission reflects the passion of its founder, to "effect a quantum change in consumer and industry behavior." That same year he published *The Organic Lawn Care Manual,* which has become something of a call to arms for people seeking to buck the chemical lawn-care trend.

Tukey, a tireless advocate with a feisty demeanor, seems to relish the fight against the Lawn-Industrial Complex. A recent annual report from Scotts said the company planned to spend $175 million on advertising alone—more than enough to make Tukey feel like David battling Goliath.

"Is it only lawn chemicals? Of course not," Tukey said. "But there are so many environmental contaminants out there that we have no control over. Car fumes, paint fumes, all the rest. How are you going to control all that? There are very few parts of that soup that you can control without extraordinary effort and expense. But you can control lawn chemicals. That's what my mission is all about. Changing the world really happens one conversation at a time—one wife changing one husband's mind, one neighbor talking to another."

Among many other projects, Tukey has produced *A Chemical Reaction,* a documentary film about Hudson, a small town in Canada that in 1991 became the first in North America to ban all lawn chemicals. Giant lawn-care companies fought the ban all the way to the Supreme Court, and lost. Twenty years later, there are more than fifty municipalities in Canada that have banned pesticides, especially on public spaces like school yards and ball fields. In 2003, Toronto began a gradual ban of nearly all lawn chemicals, including 2,4-D. Five years later, the leading lobbyist for the lawn chemical industry admitted that they'd "basically lost Canada."

It's not just Canada, either. Denmark, Norway, and Sweden have all banned 2,4-D. In January 2009, the European Union passed new pesticide laws than ban twenty-two synthetic chemicals that can cause cancer or disrupt human hormones or reproduction. Over the next five years, pesticide use will be restricted near public spaces like schools, parks, and hospitals. Wholesale aerial spraying is prohibited. (Yet the rhetorical battle continues. The chairman of the British Carrot Growers Association said that the new law would "wipe us out." Environmental groups counter that banning just 22 out of 400 harmful substances is "barely a start.")

In the United States today, some 60 million people live in communities that ban people from hanging laundry. How many communities ban the use of pesticides on lawns? Not nearly as many, it seems, as those that *require* them. A friend of mine lives in a Baltimore neighborhood that prohibits her from turning away the community's lawn-care company, even though they spray, and even though she has young children. If a neighbor can appeal to a community board to force you to take down your laundry line, can you appeal to have your neighbor stop spreading toxic chemicals on their lawn? Which is more unsettling to a community, sheets hanging out to dry or the prospect of children developing a lymphoma and women having miscarriages?

Although a few American towns—Camden, Maine; Hebron, Connecticut; Marblehead, Massachusetts—have banned spraying on public grounds, the lawn-care industry still has a stranglehold on the United States. The year they lost the Hudson battle, a group of lawn chemical companies formed a lobbying group called RISE—Responsible Industry for a Sound Environment—to represent producers and suppliers of pesticides and fertilizers in legislative fights. The group's website, PestFacts.org, states plainly that "synthetic chemicals are not a cause of cancer"; that there is "no link between spraying and asthma"; and that efforts to limit phosphorous from choking water supplies "are not based on sound science."

The website also boasts that RISE "has an excellent working relationship with the EPA and is a resource for the Agency." This, to Tukey's mind, is precisely the problem. RISE has "the express goal of not allowing what happened in Canada to happen in the United States," he said. "They can't let that happen. They were talking about building a $2 million war chest to beat back people like me."

The group's lobbyists are pushing for state "preemption" laws, which allow state governments to prohibit cities and towns from passing their own restrictions on chemicals. Like other industries, Big Lawn would prefer to keep regulation at the federal level. Industry says this is for "uniformity"; Tukey says it's because industry knows federal oversight is utterly lax—and because lobbying a few influential congressmen is a lot easier than trying to herd unruly mayors or state legislators into line.

Tukey's home state of Maine has exactly one paid state inspector for pesticides, and that, he says, is the way the lawn-care companies like it. But Rick Churchill sees Tukey's progress, from synthetics to organics, as a harbinger.

"Like me when I was young, he believed much of what the pesticide chemical manufacturers were saying about how safe their products were," Churchill said. "Who would question Monsanto? Now, of course, many of us know these companies for the dirty deeds that they continue to force on the agricultural community and, ultimately, the consumer. Of course, what I enjoy most is I believe his name is well known by the Monsantos of the world, and they would do anything to keep him quiet!"

For his part, Tukey understands that the struggle is far from over. Even his own father, a bladder cancer survivor, "hasn't given up his ChemLawn." The last time Paul went to visit, he peeked in his father's trash can and saw a handful of signs the sprayers had left, warning people to stay off the lawn. There were sixteen of them.

But rather than struggle with his father's generation, Tukey devotes his energies to his son's.

"Did I cause my son's ADHD? I'll probably never know for sure," Tukey said. "But I have to live with the thought that I probably did. The guilt sticks with me every single day. That's the saddest part of all this. The positive side is that I use my son as motivation to get the word out to others."

Firing your chemically dependent lawn-care company or replacing it with a less toxic competitor would substantially reduce your family's—and your neighbors'—exposure to synthetic chemicals.

It would also greatly reduce the volume of pollutants—both herbicides and synthetic fertilizers—that you contribute to your watershed. Both are excellent outcomes. But there's another option, one that goes beyond minimizing threats and gets into the entirely more interesting realm of restoration. There is a way to think of your yard as far more than an onerous burden that needs to be mowed every week, far more than a sponge for soaking up synthetic chemicals. There is a way to think of your yard as transformative—even, dare I say it, magical. And Doug Tallamy can show you how.

When Tallamy, the chair of the entomology department at the University of Delaware, walks around his yard, he sees things most of us would not. He can look at a black cherry tree, for example, and see the larvae of thirteen tiger swallowtail butterflies. He can spot an eastern tailed-blue butterfly depositing its eggs on arrowwood viburnum. One recent spring afternoon, as he toured me through his Pennsylvania garden, his eyes flashed over my shoulder to catch two robins mating in midflight. Later, he noticed that the leaves on a chestnut oak he had planted on a hillside had become slightly yellow. "Too much magnesium in the soil," he said. "Maybe I'm trying to force this guy into a place he doesn't want to be."

A few years ago, while on his way to give blood, Doug Tallamy stumbled on a handful of red oak acorns scattered across a parking lot. He gathered them up, put them in his lunch box, and tossed them into his backyard. Now he has thirty red oak trees. He has planted scores of other trees in his yard: sweet gums and tulip trees, white oaks and river birches, sugar maples and pignut hickories. There are chokecherries and pawpaws, dogwoods and hornbeams. And the funny thing is, it's not even trees that Tallamy is most interested in. What he's really interested in is bugs. And birds. And saving the country from its addiction to chemically saturated lawns.

Tallamy spent his childhood exploring the woods that surrounded a small lake in northern New Jersey. He would creep up to the side of a pond to watch pollywogs turning into toads. He watched them grow legs—rear first, then fore—and watched them hop up into the weeds. He learned their mating calls. Then, one day, he stared, horrified, as a bulldozer barreled over a nearby hill. What followed was a moment of terror that Tallamy still remembers as somehow threatening to himself and his world.

"In an act that has been replicated around the nation millions of times since," Tallamy wrote in his book *Bringing Nature Home*, the bulldozer "proceeded to bury the young toads and all of the other living treasures within the pond. I might have been buried too, if I hadn't given up trying to rescue the toads."

The young Tallamy saved about ten that day, but for nothing: the pond was gone, leaving nowhere for the toads to breed. As he wrote, "Within two years, a toad was a rare sight near my house; soon they were completely gone, along with the garter snakes, whose main prey they had been, and other members of the food web supported by the life in that pond. I had witnessed the local extinction of a thriving community of animals, sacrificed so that my neighbors-to-be could have an expansive lawn."

To Tallamy, the buried pond is less a sign of progress than the very symbol of human folly and ecological decay. He feels the same way about car windshields. As a boy growing up near a five-hundred-acre forest, he would see countless inchworms dangling from invisible threads attached to trees. On road trips to his parents' camp, he remembers the headlights and windshield of the family car arriving utterly splattered with insects. Nowadays, he rarely sees an inchworm, and his car always arrives as clean as it left. To Tallamy, the disappearance of forests, and of the animals and insects they once supported, are signs of an American landscape in the process of ecological collapse. The nice thing is that this is a problem that suburban home owners have the power to reverse.

Americans maintain an image of their country that is at least a hundred years out of date, Tallamy said. We still imagine that we exist along the frontier, that no matter how many subdivisions or shopping malls we build, there will always be undisturbed land left. But this image is not only obsolete, it has become ecologically dangerous.

"We're hardwired to be incredibly adaptable, but these traits may no longer be useful," Tallamy told me. "It used to be if we blew it here, we'd move over there. But now the entire planet is flooded with us. There is no place to move. We have massive populations, and there is no place to go. We're already at three or four times the earth's carrying capacity. We can't behave like we used to."

When Europeans first got to this country, they looked out on 950 million acres of virgin forest, a wilderness that scared the pants off people coming from a continent largely plowed under for hundreds of years. So what should the new country's civilized homes look like? The Founding Fathers apparently decided that at least when it came to landscaping ideas, the English had gotten things right.

"Back in Europe, only the rich people had these gardens," Tallamy said. "Then George Washington and Thomas Jefferson demonstrated their wealth by planting gardens at Monticello and Mount Vernon, and the lawn became a symbol of wealth. Today, people think, 'If I can get a big lawn, I can display my wealth.' There are plenty of home-owners' associations that have lists of plants you can use, and every one of them is an alien ornamental. The crazy thing is that gardens in Europe are using plants from North America, and vice versa. There is a lot of ignorance out there."

So here's a question: Let's say you've been using 2,4-D on your lawn for years, and—so far, at least—it has never made you sick. Leaving aside the question of how you can be sure of this, let's consider something else: How do you feel about using a product that poisons worms, which then poison the songbirds that eat them?

The toll of suburban development has been devastating to bird populations. You may see a lot of birds flying around your house, but look a bit closer: most are probably house sparrows and starlings, which are aggressive invasive species from Europe. Study the population numbers for native birds: the wood thrush is down 48 percent. The bobwhite, down 80 percent. Bobolinks, down 90 percent. Cornell's David Pimentel estimates that 72 million birds are killed each year by direct exposure to pesticides, a number that does not include baby birds that die because of the loss of a parent killed by pesticides, or birds killed by eating contaminated insects or worms. The actual number of birds killed might be closer to 150 million.

It's not just lawn chemicals that do them in, of course. Birds are also poisoned by pesticides sprayed on crops in their wintering grounds in places like Latin America, where pesticide use is up 500 percent since the 1980s; a single application can kill seven to twenty-five birds per acre. An additional 100 million are killed every year by cars. Some 100 million on top of that are killed by domestic and feral cats. A billion—a billion!—are killed by crashing into windows, transmission towers, or power lines.

In the last year, surveys done by the government, conservation groups, and citizen volunteers have found that nearly a third of the country's 800 bird species are endangered, threatened, or in serious decline. Much of this is due to habitat loss caused by suburban sprawl, as well as "a barrage of exotic forest pests and disease."

And it's not just birds. As of 2002, Delaware had lost 78 percent of its freshwater mussel species, 34 percent of its dragonflies, 20 percent of its fish species, and 31 percent of its reptiles and amphibians. Up and down the Atlantic seaboard native plant species are threatened or already extinct, as are bird species that depend on forest cover. Neotropical migrants like wood thrushes, warblers, catbirds, hawks, wrens, vireos, flycatchers, kingbirds, nightjars, swallows, tanagers, orioles—species that fly thousands of miles to Central or South America to spend the winter—have declined an average of 1 percent a year since 1966. Overall, that's 50 percent reduction in just over 50 years.

"It is curious that the news media have drawn our attention to the loss of tropical forests yet have been silent when it comes to how we have devastated our own forests here in the temperate zone," Tallamy has written. "Only 15 percent of the Amazonian basin has been logged, whereas over 70 percent of the forests along our eastern seaboard are gone. We have reduced the enormous land mass that, over millions of years, created the rich biodiversity we can still see today in this country to tiny island habitats. And therein lies the problem."

But Tallamy is far more than an academic doomsayer. In mid-Atlantic gardening circles, he has become something of a prophet, his message freighted with both gloom and promise. And it is the latter, the promise of ecological renewal, that Tallamy most wants the lawn mowers of the world to understand.

While the mid-Atlantic may not host the vast national parks of the American West, it could—given the slightest shift in human behavior—return to its status as one of the country's richest ecosystems. All it will take is a shift in the way we think about our lawns and gardens.

Tallamy's vision is based on three simple ideas. If you want more birds, you need more native insects. If you want more native insects, you need more native plants. And if you want more native plants, you need to get rid of—or at least shrink—your lawn. Contrary to our received wisdom—reinforced by five decades of marketing from the lawn chemical industry—chemically treated grass is not the apotheosis of suburban living. As strange as it might feel to lure insects into your garden, if your garden is balanced, its overall health will be fine. The bugs will eat only a small percentage of your leaves, and their hunger will be more than compensated for by the number of birds they attract. A garden with no insects is not a sign of perfection. It is a sign of sterility—and thus, most likely, a sign of chemical toxicity. To Doug Tallamy, such an arrangement "is good for the pesticide industry, but little else."

The impulse to rid a landscape of all insects strikes Tallamy as especially pathological. He once fell into a conversation with a biology professor, and the talk turned to bugs—and how best to exterminate them.

All of them.

Tallamy was stunned. As a field scientist, he mourns the loss of field training among his colleagues. To him, learning how to kill insects in a laboratory is akin to learning the physics of bomb making: it may be a useful scientific challenge, but what it fails to consider—the collateral damage to entire ecosystems—is vast. A land without insects may seem like a gardener's dream. But a land without insects is also a land empty of higher species.

I asked Tallamy about the theory that humans tend to chop down trees and lay out lawns because they remind some ancient part of our brains of open savannahs: we plant lawns so we can see the lions before they see us. Tallamy thinks this idea is reasonable, if terribly outdated. "We humans are obsessed with collecting rare things," Tallamy told me. "We collect stamps, or the biggest ball of yarn in South Dakota. We want to get the plant that our neighbor doesn't have. If the plant is common, like a jack-in-the-pulpit, we think, 'Oh, that's no good.' That rules out native plants right from the beginning."

A couple of years ago, I began to take Tallamy's argument seriously. I went around looking to buy native plants and put them in my yard. This wasn't that easy. I'd drive out to garden stores and nurseries and look around. The places were filled with lovely shrubs and trees and flowers, most of which were never meant to grow in Maryland. Japanese maple trees. Forsythia. English ivy. Pachysandra. Native plants are not typically labeled as such, and neither are exotics. Or invasives. I've seen purple loosestrife for sale, despite its being an absolute scourge in East Coast river systems. When I asked one nurseryman where he kept his black cherry trees and his elderberries, he looked at me liked I'd asked to buy poison ivy.

"Oh, those are common," he said. "I don't know who would carry those."

That's when I realized just how challenging this might be. If a plant is "common"—that is to say, easy to find in Maryland—it is, by definition, "undesirable."

Native trees, Tallamy has since found, support thirty-five times more caterpillar biomass than aliens. In other words, there is thirty-five times more bird food on trees that are native to a place than on trees imported from somewhere else. The trouble is, home owners and real estate developers do not take caterpillars into account when they are buying or building homes. In the sprawling suburbs of the mid-Atlantic states, for example, developers often begin a project by stripping the ground of all vegetation, trees and shrubs alike. They scrape away the topsoil, build their houses, lay down chemically dependent carpets of sod, and plant inexpensive ornamental trees—typically nonnative trees like Bradford pears and Norway maples, which look pretty but have no ecological value whatsoever. Because the grass and the trees evolved somewhere else, insects native to, say, Pennsylvania or Delaware cannot eat their leaves. That might seem like a good thing. But it also means that these numberless acres of lawns and ornamental trees provide no food for birds. What this means for suburban gardens, of course, is that planting exotic ornamental shrubs or trees will virtually ensure that birds stay away. If enough ground is planted this way, birds won't just stay away. They will starve to death.

By the time you wake up to morning birdsong, Tallamy says, migratory songbirds may have just landed after flying three hundred miles. They have descended into your trees exhausted, desperate for something to eat. What, exactly, do they find? All too frequently, they see ornamental trees that bear none of the insects they need to survive—and acres and acres of chemically treated grass.

"When birds are flying over a vast swath of suburbia, they can't just say, 'I'll just fly another one hundred miles,'" Tallamy remarked. "If they land on a patch of Bradford pears, they're out of luck." Despite the promising name, Bradford pears bear no true fruit, and are an unsatisfying food source for birds.

Tallamy and his wife, Cindy, used to live near West Chester, Pennsylvania, a bedroom community between Wilmington and Philadelphia. In 2000, longing for more space, they bought ten acres of neglected farmland in rural Oxford, Pennsylvania, in the extreme southeastern corner of the state. A centuries-old cattle farm, the land had been mowed or grazed for years, but for more than a decade it had been left entirely alone. By the time the Tallamys moved in, the place was utterly overgrown with aggressive alien species, most of them now anchored with thick trunks and substantial root stocks. Vines of Oriental bittersweet six inches thick had climbed to the tops of most native trees. The autumn olive trees were growing four feet a year.

When Tallamy looks at a landscape, his eyes take in far more than just plants. He sees relationships: between plants and soil, between plants and bugs, between plants and birds. And what he saw in those early years was striking.

"When we walked around the place in those early years," he recalled, "we were just like the birds: we looked for leaves that had insect damage. Nothing had been eaten out of the ornamentals. We already know that insects evolve with plants. It takes a long evolutionary time for an insect to switch plants."

Armed with bow saws and powerful loppers, Doug and Cindy set out to "take the land back." Fearful of destroying the few native trees and plants that might be hiding under the jungle of invasives, the Tallamys decided not to mow everything down with a brush hog, and they had no interest in mass spraying. Instead, they used a small paintbrush to swipe newly cut exotics with the poison glyphosate, sold by Monsanto as Roundup. Not exactly organic, but, Tallamy says, without a dab of poison, there's no getting rid of things like ailanthus. Even if you cut the so-called Chinese tree of heaven to the ground, it will return to haunt you year after year after year.

The Tallamys ripped out impenetrably deep thickets of multiflora roses and replaced them with native trees and shrubs. Out came the Oriental bittersweet. And the autumn olive. And the Japanese honeysuckle. In went the black cherries and the white oaks, the pignut and mockernut hickories, the poplars and sycamores and native willows. The choices they made were for more than aesthetics. The Tallamys plant oaks because individual trees can support hundreds of species of insects, including moths, butterflies, and inchworms. Ditto for black cherries, which also provide a critical crop of fruit for migrating birds.

This done, they replenished their topsoil. They built birdhouses to attract Carolina wrens. They planted blackberries to draw yellow-breasted chats. For borders with the road and their neighbors, they planted white pines. The results have been startling. A couple of years ago, a blue grosbeak nested on their property, probably for the first time since before the land was first farmed, centuries ago.

The Tallamys' property is bordered on one side by a huge race-horse training facility and on the other by a private home set on a piece of property about the same size as the Tallamys'. The horse business maintains a half-mile track with an expansive infield the owner—holding with tradition—keeps meticulously mowed. To Tallamy's eyes, the infield, with a slight attitude adjustment, could be allowed to revert to a native grassland meadow that would provide nesting areas and food for bobwhites, meadowlarks, bobolinks, and grasshopper sparrows—all without hampering the horses in the least. To Tallamy, a small shift like this can have major ecological effects.

On the other side, Tallamy's neighbor Sam provides what Tallamy called a "counterpoint." Sam also owns ten acres, but he maintains it "like a golf course, with tractors, commercial lawn mowers, weed whackers, and leaf blowers roaring more days than not." Sam's yard is planted almost completely with alien species, a lot of them put in by "professionals."

When Sam asked Tallamy why he didn't just mow his meadow, Tallamy was taken aback. The field provided refuge for hundreds of toads; food and nesting cover for indigo buntings, bluebirds, blue grosbeaks, and field, grasshopper, and song sparrows; a courtship arena for woodcocks in the early spring; and a place for foxes and great horned owls to catch rabbits, voles, and mice in the winter. "The unmowed field also keeps the land surface from becoming so compacted by heavy machinery that it cannot absorb the hard rains that recharge our water table before the water runs off to the nearest creek," Tallamy wrote in *Bringing Nature Home*. "Realizing that Sam and most of the people in this country, through no fault of their own, are now so divorced from nature in their education and their everyday lives that they do not know what nature is, why we need it, or in what ways it is wonderful left me uncharacteristically tongue-tied."

Doug Tallamy's prescription for suburbia is simple in practice but profound in implication. Rip up your lawn. Instead of obsessing over your grass—or, worse, dousing it with chemicals—consider planting native things that will not only look lovely but will also make your garden a haven for caterpillars, butterflies, and birds. In the mid-Atlantic region, this means planting things like swamp milkweed and butterfly weed and buttonbush and joe-pye weed (*Eupatorium*) and rudbeckia species like black-eyed Susans (not for nothing Maryland's state flower). Tallamy's book has a chapter called "What Does Bird Food Look Like?" with eye-popping photos of the caterpillars—and birds—he has discovered on his native shrubs and trees. But Tallamy's garden is only incidentally beautiful. What he's really after—and what he wants other suburbanites to understand—is that replacing toxic and barren lawns with native plants is nothing short of "a grassroots solution to the extinction crisis." We don't need the government, we don't need to keep our fingers crossed that conservation organizations will do the work for us. We can do it ourselves.

At the University of Delaware, Tallamy has joined a team planting native species all over the campus—no 2,4-D or mowing required. One of Jerry Kauffman's graduate students recently ripped up an old storm drain and turned it into a rain garden that helps purify water pouring off a university parking lot. A campus "lepidoptera garden" will consist entirely of native plants that offer food and shelter to native caterpillars, butterflies, and moths, which, like native songbirds, have been in significant decline. Beyond their function as a habitat, the gardens also offer vivid examples of smart landscaping to the countless students and families walking across the campus every day. Instead of looking out over acres of nonnative grass that requires regular doses of petrochemicals, they will see plants and trees that are as native to Delaware as the students themselves. Perhaps, in getting to know their native landscape a bit more intimately, these students will come to find joy in their surroundings, rather than fear.

And me? Last spring, I decided to take Doug Tallamy's words to heart. I ripped up a good 20 percent of my eighth of an acre and planted two large flower gardens, two substantial sets of flowering shrubs, and seven raised-bed vegetable gardens. The benefits are manifest. My daughter, Annalisa, now five, helps me pick eggplants and tomatillos and okra and Swiss chard. My son, Steedman, now nine, can identify not only monarchs and tiger swallowtails but which plants they like to eat. Why? Because last year the butterflies were not here, and this year they are. Because this year we ripped up the grass, which monarch caterpillars can't eat, and planted milkweed and butterfly weed, which they can. Simple as that. Toxic herbicides and petrochemical fertilizers? Who needs them? Milkweed and joe-pye weed were born to grow here. All you have to do is plant them, and wait for the butterflies.

Epilogue: **What's Next**

Given the world we currently inhabit, it's easy to feel overwhelmed. Synthetic chemicals are literally everywhere; they seem hard to name, harder to avoid, and even harder to replace. So what are we supposed to do? What choices should we make? Where do we even begin?

Some answers to these questions are simple, and some are quite complex. Some require only small changes, and some could, if taken seriously, lead to deep adjustments in our personal lives and the way we think about entire systems. Changing brands of toothpaste or throwing out the roach spray is easy. Learning how to paint our houses—and our faces—with products that won't harm our health is a bit harder. Firing the lawn service and planting native species—an action that would improve the health of your family *and* your watershed—requires an even bigger shift. Getting toxins out of our food? Making our rivers clean enough to swim in? How far do you want to go?

As with any investigation into the subtleties of the human body or the natural environment, every thread is connected to every other thread. Pull one, and you tug at the whole—this is the lesson Swedes learned when they banned toxic flame retardants and saw breast cancer rates drop by a third. Such connections are everywhere. What's the link between a light bulb and a can of tuna, for instance? It turns out that much of the mercury that ends up in the fish we eat comes from emissions from coal-fired power plants. So if you want to reduce your body's mercury levels, you might think about switching to wind power, and convincing your neighbors to do the same. Once you learn that the packaging of greasy prepared foods contains Teflon chemicals, you have one more compelling reason to eat better. And so on.

"The hopeful part of all this is not that we have to give things up," Russell Libby, the organic farmer in Maine, told me. "It's that if we can make changes and address our prime exposure pathways, we can make a big difference fairly quickly. Twenty or thirty years ago, the first organic farmers were operating at a really small scale. There were a lot of failures. People would lose a crop because they didn't yet know how to do it successfully without pesticides. By the time we got to ten years ago, there were at least some people who could grow every crop without pesticides. Now all those practices are starting to cut across into conventional agriculture."

The paths to making similar changes with other products, Libby said, are parallel: you develop alternative technologies or product designs; and you push policy that encourages—or forces—the development of these alternatives. Until both of these things happen in tandem, we will remain stuck with what we have.

"Can an individual solve this? Only a little bit," Libby said. "You can make decisions at a few key points in your life. You can make a difference each time you buy a car. You can have an impact on your food and your ongoing consumer purchases. We can turn the thermostat up and down, make good choices in our furniture, our toys. But the major pathways are systemic issues that have to be addressed as a country, and that's really the challenge. When people have a severe illness, getting politically active is not the first thing that you think of. It's all about 'How do I get better?' Once

you're better, often you've been through such a major challenge—
your household finances are severely impacted, you're trying to
find a pathway back to normal—and yet that might be exactly the
time when 'normal' might need to include the bigger system.
That's the heart of the policy challenge. We haven't really made
this a public policy discussion yet."

So, Libby said, we need to learn to shop smarter. But we also
need to vote better. Only when we do both will we decrease our
exposure to toxic chemicals. Thankfully, there are pioneers out
there ready to take the lead.

Wendy Gordon hardly looks like the revolutionary type. Her neat
blond hair and matching coral sweater and skirt might be more
suited to the produce aisle at a Whole Foods than the Maine
Agricultural Trades Show. But in the push for safer consumer
products, Gordon is firmly on the front lines. She grew up along
the Hudson River, in Englewood, New Jersey, and Ulster Park,
New York. Like anyone else coming of age in that region in the
1970s, she thought of the Hudson as a chemical sewer, all but dead
because of the PCBs and other toxins dumped into it since the
1950s.

For her geology degree at Princeton, she studied a groundwater contamination case at an IBM facility in central New Jersey.
She went on to get a master's degree in environmental health sciences from Harvard's School of Public Health. She did an intern-
ship with the Natural Resources Defense Council, and ended up
staying a dozen years. She worked mostly on the staff of the Toxics
Project, using scientific studies to support legal cases against polluters or changes in public policy. It was at the NRDC that
Gordon learned about federal laws regulating everything from in-
secticides to Superfund sites—and saw just how hard it was to
change the way people think about toxic chemicals.

Her focus was on the exposure of infants and toddlers to toxic
pesticides found on fruits and vegetables. When she started, she
noticed that most studies done on the health effects of pesticides

had been conducted on healthy twenty-one-year-old males, a group that seemed to offer little insight into the heath of children under six. The vulnerability of the younger subpopulation, she realized, had never been taken into account. Armed with a series of new studies, NRDC found that pesticides that were considered safe for adults were unsafe for infants; and not long after this, the National Academy of Sciences confirmed these results.

In 1989, the NRDC crystallized these findings in a report called "Intolerable Risk," which warned about the health risks to children from eating fruits and vegetables contaminated by agricultural chemicals. Of particular worry was the chemical daminozide, more commonly known by Uniroyal Chemical's trade name Alar; the compound was sprayed on apple orchards so that the fruit would all ripen at the same time. In 1984 and again in 1987, the EPA had listed Alar as a probable carcinogen, and in 1986 the American Academy of Pediatrics had urged the EPA to ban it. Even before the study, six national grocery chains and nine major food processors had stopped accepting apples treated with Alar. Maine and Massachusetts banned it outright. But the NRDC study struck a chord because it made the public connect the dots between a chemical, Alar, and the potential for cancer in children who ate, of all things, applesauce.

Not long after the study was released, 60 Minutes ran a story on the report, and 40 million viewers simultaneously panicked. The outcry became so intense that Uniroyal was forced to pull the chemical off the market.

The Alar story broke full force just as Gordon was going on maternity leave, and for a young mother, the moment opened her to new ways of thinking about toxins and health. The NRDC had already begun working on ways to "detoxify our kids' food supply," Gordon said, but the more they pressed the idea of toxic produce as an environmental issue, the more parents seemed to tune out.

"We couldn't get parents worked up about laws," Gordon told me. "They just wanted to know what apples to buy." By the time she got back to work, Gordon had decided to organize "an upset-parent population." She and Meryl Streep, who had been an

NRDC spokesperson during the Alar flap, cofounded an organization called Mothers and Others for a Livable Planet, in order to encourage consumer interest in organic food and sustainable agriculture. This led, in 1994, to the creation of a bimonthly consumer newsletter called the *Green Guide*, which became a website in 2002, aimed mostly at women between the ages of twenty-five and fifty-five. The site became so popular that in 2007, Gordon sold the newsletter and the website to *National Geographic*, which launched the first issue of the magazine in the spring of 2008.

Just as Big Chemical has taken its cues from Big Tobacco, so have advocates of nontoxic consumer products taken their cues from the organic food movement: information is everything. What started out as an environmental standard, concerned with reducing the use of "toxic inputs" on farms, quickly evolved into a kind of seal of approval, a social standard. To Gordon, the "organic" label that now appears on everything from fresh lettuce to children's cereal has created the ideal way to change consumer behavior: give people information, and give them a choice. Looking over a dizzying array of options in the supermarket, the little green USDA ORGANIC seal has come to mean something beyond whether or not the farmer sprays his fields with 2,4-D. She'd like to see a similar shift on all products.

"When you're a mom, you just don't want to be constantly reminded of all the things that are bad for you and your babies," Gordon said. "Nobody's all that interested in saving the planet. All they care about is taking care of their kids. There are all these talking heads saying, 'Yes, it's a carcinogen [or] No, it's not a carcinogen.' The bad guys could tie the good guys up in endless arguments. The *Green Guide* started as me trying to be a smart mama and trying to help other people be smart mamas. We were just trying to cut through the noise to make smart choices. If you don't know the impact of something, why not avoid it? If you're at the store and you don't know what's in something, why take the risk?"

The magazine found its voice when consumers began wringing their hands over the safety of milk produced by cows treated with bovine growth hormone. Rather than join the debate, Gordon's group published its "mother's milk" list, simply showing parents a roster of dairies that refused to use growth hormones. It became the most popular thing the magazine had ever done.

In a column in the magazine's inaugural issue called "Just Ask," a mother wrote that she had read recent news stories about lead in children's toys and wanted to know how safe it was for her twin three-year-olds to use "play" cosmetics. The magazine answered with brio, saying, "No mom wants an innocent game of dress-up to turn her children into guinea pigs for the cosmetics trade," and recommended that she look for products with plant-based dyes and petroleum-free alternatives like beeswax, castor oil, or shea butter. Synthetic FD&C dyes and petroleum by-products are possible sources of lead, the column noted, and glitter sticks may also contain parabens and glycol ethers. It plugged the Campaign for Safe Cosmetics report on lead in lipstick and warned that nail products contain toluene, dibutyl phthalate, and formaldehyde, all three of which are on California's list of chemicals known to cause cancer and/or reproductive disorders.

Gordon considers herself a part of the "third wave" of environmental work. The first wave, exemplified by the likes of John Muir and Theodore Roosevelt, were "conservationists," fighting to keep large parcels of land off-limits to human development. The second wave, heralded by the publication of *Silent Spring* and enforced with legislative breakthroughs like the formation of the EPA and the Clean Air Act, established benchmarks on industrial pollution. The third wave, Gordon says, is much more about the individual consumer, about trying to change corporate behavior not through politics, which industries control, but through consumer pressure. The key is getting consumers to think harder about what they buy.

"We can't keep chasing the end-of-the-pipe polluter," she said. "Waste is very expensive. We had the underlying theory that an informed consumer would make wise decisions. We're looking much more closely at economical consumption and sustainable production. There are lots of parallels between the industrial polluter and the consumer of their products. It doesn't take too large a group of consumers to get producers to change their products. This is much more efficient than trying to work the legislative process, which can take years."

Indeed, as the editor of what is, ultimately, a consumer magazine, Gordon has to tread a fine line between cheering smart consumerism without cheering consumerism itself. The *Green Guide* calls itself "The Resource for Consuming Wisely," which hardly sounds like a Marxist manifesto. The magazine is printed on Forest Stewardship Council certified paper, so its pulp comes from well-managed forests. Ten percent of its paper content comes from recycled postconsumer waste. Its ink has no heavy metals and is derived from soy, corn, and linseed.

Gordon knows that her audience—women of childbearing age—is a powerful group, comprising not only voters but also shoppers. And it is her bet that getting them to vote with their buying habits is the most efficient way to get companies to clean up their product lines.

"No retailer is ever going to try to fight mothers," Gordon said. "It's a real easy choice. It's real easy to go shop down the street."

This is a lesson Martin Wolf has taken to heart at Seventh Generation, and it a lesson that is even beginning to trickle down into university chemistry labs. Although few American universities currently require chemistry students to demonstrate knowledge of the health and environmental consequences of the chemicals they create, there is a burgeoning field of "green chemistry," which seeks to reduce or eliminate the use of toxic synthetic chemicals.

"Of all the chemical products and processes currently in use, only 10 percent are benign and nontoxic; the other 90 percent have hazardous consequences," says John Warner, the director of the world's first green chemistry PhD program, at the University of Massachusetts–Lowell (and the coauthor, with Paul Anastas, of *Green Chemistry: Theory and Practice*). Of those toxic materials, Warner estimates that 25 percent could be replaced, right away, with safer alternatives.

And the other 75 percent?

"Today we do not have solutions to provide to industry," Anastas says. "The current way we have chemistry set up does not even touch these needs. People say there's nothing new to invent. This is a whole new area for innovation, for creativity, for cutting-edge technology. It's a huge opportunity."

In recent years, some industries have started to acknowledge this. In 1999, Baxter International, one of the largest makers of in-travenous hospital products, agreed to stop using PVC in its products after a group of Catholic and labor union shareholders threatened to dump their shares of Baxter stock. More recently, the health care giant Kaiser Permanente, a company with $38 billion in annual revenue, put out a call for PVC-free products, and manufacturers perked up in a hurry. Kaiser now buys carpets made by Tandus, of Dalton, Georgia, which uses a nontoxic polymer called polyvinyl butyral, or PVB. It buys nontoxic intravenous bags from Baxter and gloves made with a latex substitute from Cardinal Health.

Since 1996, the EPA has been handing out Presidential Green Chemistry Challenge Awards to companies that invent environmentally benign products. Recent winners have included the makers of soy-based toners for laser printers, formaldehyde-free plywood, and ingredients for paints that do not contain volatile organic compounds. In 2003, the award was won by William McDonough, who helped Shaw Industries, a rug-manufacturing company owned by Warren Buffett, figure out how to make carpets without the use of PVC. McDonough is something of a green-design visionary; he wants people to reimagine what it means to be a "consumer," since everything we "consume" (except for a bit of water and food) is designed to be discarded. Changing the way products are designed, to make them both recyclable and chemically benign, will be one of the century's great industrial challenges.

"In a world where designs are unintelligent and destructive, regulations can reduce immediate deleterious effects. But ultimately a regulation is a signal of design failure," McDonough writes in his book *Cradle to Cradle*. "In fact, it is what we call a *license to harm:* a permit issued by a government to an industry so that it may dispense sickness, destruction, and death at an 'acceptable' rate. But as we shall see, good design can require no regulation at all." McDonough's book is printed on a substance made from PET, a nontoxic synthetic. Books, he notes, are made of a number of materials that can be toxic, including the ink and the glues used to bind them. If books were printed on PET, consumers could ship them back to the manufacturer to be melted down and made into new books. Patagonia, using a design McDonough developed, accepts old boxer shorts and long underwear for recycling into new fibers. The mailbox, in McDonough's view, can become the new garbage can. "It means nothing has to be junked," McDonough said. "It means life after life."

Last year, in an effort to stem the tidal wave of discarded plastics, France put a tax on things like nonrecyclable utensils and plates, and may take similar actions on everything from televisions to refrigerators. In California, Hewlett-Packard recycles some 1.5 million pounds of electronics every month, an honorable act that nonetheless barely dents the two million *tons* of electronics that get dumped into American landfills every year.

DuPont is working to turn fermented corn sugar into a polymer that will replace petroleum products in paint. Cargill is making a compound from vegetable oil that may replace polyurethane foam in seat cushions. In Maine, a recent grant is helping the state's distressed potato farmers in Aroostook County turn their crops into biodegradable plastic, a move that Mike Belliveau, one of the principals in the state's body burden study, considers a good omen. "One advantage we have over the climate-change crowd is that the solutions to climate change require fundamental changes to the whole economy," Belliveau told me. "At the moment, our economy is almost entirely based on the burning of fossil fuels. But our economy does not inherently require inherently dangerous chemicals. The coal, oil, and gas industries are resistant to climate-change legislation. On this issue, it's really only the chemical industry.

"Most of the user industries," Belliveau pointed out, "are already seeing the writing on the wall. Electronics and cosmetics industries are already looking into alternative materials. Apple and Dell are competing to see who can produce machines that are free of halogens like chlorine, bromine, and fluorine. There are probably five or ten thousand halogenated compounds out there. The DuPonts, the Dows—these companies are wedded to these compounds. If you ban one, they'll put out five more. Most of these chemicals have never been tested for public health, and never will be, because there's no testing required. But now leading-edge companies are saying we want to get out of halogens altogether. That's a huge leap."

And the truth is, just as toxins can accumulate and spread around the world, so, too, can information—and, occasionally, political action. The breast milk studies coming out of Sweden and Texas, for example, swirled around the globe and began catching the attention of lawmakers in Europe and—in some small pockets, at least—the United States. One person who took notice was a young state legislator from Maine named Hannah Pingree, who had recently been elected to represent the people living on a handful of islands off the Atlantic coast.

People in Maine tend to be self-sufficient and suspicious of "people from away." Clothing runs to flannel shirts, wool sweaters, and Carhartt jackets. In a pub near Augusta one frozen January night, I fell into a conversation with a young state legislator with little good to say about a wealthy state to the south; he referred to it only as "Messy Two Shits." Driving to the statehouse, I spotted a bumper sticker with a cartoon of a man pointing to a smear of lipstick on his hindquarters. These are the kinds of independent spirits Pingree represents.

Pingree grew up on an island called North Haven, a fishing community with 380 year-round residents. There is no bridge to the mainland, only a ferry that runs three times a day. There are no industrial polluters on North Haven, no chemical factories, no paper mills. She still lives there.

Around the time Sweden was banning flame retardants, Pingree began thinking about pushing the state to consider what she called "a more rational chemical policy" and to convince her fellow lawmakers to make Maine one of the first states in the country to ban the use of octa and penta. But the more she learned about toxic chemicals, the clearer it became that the issue went far beyond the banning of a couple of flame retardant chemicals.

It was about then that the body burden people called, asking her to take part. Pingree jumped at the chance.

Like Lauralee Raymond, Pingree felt sure her results would be pretty clean. She, too, was young, fit, and rural-born. She had recently gotten married and was beginning to think seriously about having a family. What was the worst that could happen? If her body turned out to be contaminated, she could use the information to push for stricter chemical regulations. But the more she thought about it, the more a creeping anxiety began doing battle with her political will. Why on earth, she wondered, did she want to learn this information? It was like visiting a palm reader: Do you really, in your heart of hearts, want to know all your hidden secrets? Nonetheless, she took a deep breath and donated samples of her blood, hair, and urine. And then she waited.

On a Sunday night in January, Pingree nervously dialed Dr. Rick Donahue, the study's principal investigator.

The results hit her like a fist.

Here she was, in her early thirties, with the study's second-highest levels of phthalates and mercury levels above the standard for protecting a developing fetus from "subtle but permanent brain damage." Not only that, but she had traces of nineteen different flame retardants in her body. Nineteen.

"I was personally outraged by all this stuff," Pingree told me. "It made me incredibly passionate about it from a policy point of view. I was very personally impacted. After I hung up the phone, I was no longer going to be a casual political supporter of a more rational chemical policy. I was now a convert who couldn't stop talking about my results."

The night she learned her results, Pingree sat down and drafted an email to every woman she knew.

"Hey, Ladies, I am participating in a study on toxics in the human body and was recently tested for a number of the major toxics," she wrote. She described her high mercury levels, warned her friends that the neurotoxin was particularly dangerous for developing fetuses, and listed the fish—tuna, swordfish, king mackerel—that pregnant women ought to avoid eating. Especially for women living along the coast of Maine, contaminated seafood is a major concern.

Pingree also warned her friends about phthalates, which, she confessed, she had never heard of before the study. She had never been a big user of cosmetics, a fact she attributes to growing up with a mother who was "a big hippie." Nonetheless, she listed the likely sources for her high phthalate exposure—perfume, nail polish, face cream, shampoo—as well as the plastics used in things like plastic water bottles and children's toys. Check the labels on all personal care products, she recommended. If they list "fragrance," chances are they contain phthalates. "Those you smear into your skin are obviously the worst.

"OK, I could go on and on about all the things I learned, but these things seemed like big ones for women," she wrote. "We consume so much of this stuff without any notice. This is just the tip of the iceberg. This info is especially powerful to folks with infants, kids, and who are likely to have kids in the next few years. I tell you this because I care about you and want you and your family to be healthy. Keep in touch, Hannah."

By now, the young legislator had become someone who could do more than write emails to her friends. At the tender age of thirty, she was now the state's house majority leader. Suddenly, at least when it came to chemical exposure, she was in a position to make some real changes.

Over the years, the American government has come up with different methods to regulate toxic chemicals. Some have been quite effective, some less so. Many are shockingly out of date. Even the weakest are under constant pressure from industry, which sees in regulation an impediment to production and, therefore, profit. How we think about the relationship between business and the government we elect to oversee it has a great deal to do with the way we think about chemicals, our health, and our environment.

Although Rachel Carson's *Silent Spring* was published to enormous acclaim in 1962, it wasn't until 1976 that Congress enacted the Toxic Substances Control Act (commonly pronounced "Tosca"), giving the EPA oversight over the testing, risk assessment, and regulation of industrial chemicals. The intent was to control chemicals that present "an unreasonable risk of injury."

Yet the policy has been badly broken from the beginning.

For starters, it grandfathered in some 62,000 chemicals already in wide commercial use. There were few questions asked, and no requirement that companies provide information about a chemical's potential harm. Thirty years later, in 1997, the Environmental Defense Fund found that basic data was missing from 71 percent of the bestselling chemicals in the United States and that the EPA had not done even perfunctory screening, let alone in-depth toxicity reports. When the EPA did its own study, it found that only 7 percent of so-called high-production-volume chemicals had a full set of data on properties and effects. Today, 99 percent of chemicals in current use are still the ones that were grandfathered in back in 1976. And since then, the EPA has used its authority to require testing for fewer than 200 of those initial 62,000 chemicals.

For another thing, under TSCA, the burden of collecting data on a chemical's dangers is left to the EPA, rather than the company that makes the substance. With limited government resources, of course, this work never approaches completion. Even the federal Government Accountability Office acknowledges that "the legal standards for demonstrating unreasonable risk are so high that they have generally discouraged EPA from using its authorities to ban or restrict the manufacture or use of existing chemicals."

Finally, the EPA is hamstrung by a requirement that it not "im-

pede unduly or create unnecessary economic barriers to technological innovation." In other words, regulators are forbidden from
acting unless they can demonstrate that restricting a chemical will
benefit society more than it will hurt a business. Today, fully 85
percent of the chemicals for which companies file for EPA review
contain virtually no health data. And even when companies do
volunteer their research to the EPA, nine times out of ten they are
marked "confidential." In 1998, 40 percent of the substantial-risk
notices filed by manufacturers asserted that the very *identities* of
the chemicals were confidential. Lynn Goldman, who served as
assistant administrator for toxic substances at the EPA from 1993
to 1998, told the journalist Mark Schapiro that she and her colleagues knew TSCA had no teeth. "There were thousands of
chemicals out there, and we didn't know what they were,"
Goldman said. "We weren't able to get the data, weren't able to assess the risks, nothing." Goldman recalled a party in Washington to
commemorate TSCA's twentieth anniversary. "Someone from the
chemical industry got up to salute TSCA and said, 'This is the perfect statute. I wish every law could be like TSCA.'"

So: companies like TSCA because the EPA doesn't make them
work too hard to prove a chemical's safety. The government lives
with TSCA because its own shrinking budget doesn't allow agents
the money to do their jobs effectively. Chemical regulation, in
other words, has become a Catch-22: the EPA lacks the power to
request data on chemicals in order to determine if they can cause
harm, and it can't make a risk assessment without these data. So
no data are provided, no risk assessments are made, and chemicals
keep flooding the market.

The upshot of all this is that companies want to maximize
profit and are thus unlikely to broadcast that they are making
chemicals or products that might be considered hazardous and,
thus, attract government regulation, bad publicity, or litigation.
Companies, in other words, have a clear incentive not to look for
trouble. If there is no worrisome data about a chemical, there is no
need to do a risk assessment. If there is no risk assessment, there is
no need to report ominous findings.

Twenty years ago, as a young newspaper journalist, I used to tromp down to the local EPA office and comb through old metal cabinets full of manila files, looking for how many tons of toluene or heptachlor or heavy metals had been released into the environment by the industries around Atlanta. (This is the stuff a senior colleague called, collectively, Methyl Ethyl Badshit.) The files were only a couple years old at the time; Congress had established the Toxics Release Inventory program, known as the TRI, in 1986, just two years after the chemical disaster at a Union Carbide pesticide plant in Bhopal, India, that killed thousands of people as they slept in their homes.

Before TRI, no one—not workers, not ordinary citizens—had the right to know anything about the toxic chemicals that were pouring out of the nation's factories. Chemical releases were considered "privileged information" or "trade secrets." After Bhopal, some things began to change, including, it turns out, the willingness of women of childbearing age to stand up to the chemical industry. "We are not expendable," said Rashida Bee, accepting the 2004 Goldman Environmental Prize for environmental justice work in Bhopal. "We are not flowers offered at the altar of profit and power. We are dancing flames committed to conquering darkness and to challenging those who threaten the planet and the magic and mystery of life."

In the United States, TRI was established as part of an emergency-planning and right-to-know law that required companies to reveal which of about 650 chemicals they were releasing into the air and water, dumping into landfills, or burning in incinerators.

And what a lot they were—and are—pumping and dumping. In 2007, industries reporting to TRI claimed that they had released 4.1 billion pounds of toxic chemicals into the air and water in the past year. Since not all industries and facilities report to TRI, these numbers are surely a low estimate, especially since the Bush administration relaxed reporting regulations for smaller polluters, regulations that were reversed soon after President Obama took office.

In 2004, the numbers were similar. In that year, according to an analysis done by the U.S. Public Interest Research Group, American companies, led by the chemical and paper industries, released more than 70 million pounds of recognized carcinogens into the air and water. That same year, companies released 96 million pounds of air and water emissions of chemicals linked to developmental problems such as birth defects and learning disabilities, almost 38 million pounds of chemicals linked to reproductive disorders, and more than 826 million pounds of suspected neurological toxins. And these, of course, are just the by-products of production. Many times more of these toxic chemicals went into consumer products and were sold to unsuspecting consumers—who ended up using and disposing of them with no regulation whatsoever.

It might not surprise you to learn that if you look at a map of the United States overlaid with the places that see the greatest quantity of toxic chemical releases, you'll find them clumped in the poorest and most rural parts of the country, most of them in the South. Texas, South Carolina, Louisiana, Tennessee, Alabama, and Florida—they all routinely rank highest for releases of carcinogens. In fact, in 2004 a quarter of all air and water releases occurred in just twenty counties, and four counties in Texas—Harris, Galveston, Brazoria, and Jefferson—ranked in the top five counties nationwide for most toxic emissions. Harris and Galveston combined for nearly 5 million pounds of releases alone, almost as much as the next eight counties combined. Two-thirds of the dioxins released came from two Dow Chemical plants in Freeport, Texas, and Plaquemine, Louisiana.

But it might surprise you to learn that the EPA has a full set of toxicity data on just 7 percent of these so-called high-volume industrial chemicals.

Ten years ago, the EPA estimated the cost of a full screening test for reproductive and developmental toxicity at about $205,000 per chemical. Chemical companies balk at this cost—despite the fact that in 2004, the chemical industry made a profit of $17 billion.

In 2006, the European Union decided to take a dramatically different approach to the regulation of toxic chemicals—though not without a fight, and not without a major push from the American chemical industry to keep things the way they'd always been. It adopted the so-called REACH legislation (for registration, evaluation, authorization, and restriction of chemical substances), which requires the registration of all chemicals produced or imported in volumes over one metric ton with the Helsinki-based European Chemicals Agency. It affects over 30,000 chemicals currently in use. Under the new law, chemicals known to cause cancer, alter genes, or affect fertility must be the first removed from the market, unless their makers can prove that they can be "adequately controlled." Beyond assessing raw chemicals, REACH also regulates "downstream" manufacturers, forcing everyone from toy makers to manufacturers of plywood to find out and report on the chemicals in their products and what effects they have on health and the environment.

REACH also dramatically reduced the amount of information companies can consider a "trade secret" and got rid of the distinction between "grandfathered" and "new" chemicals; under the new policy, everything has to be registered. REACH also limits the use of chemicals shown to be unusually persistent, as well as those that bioaccumulate in the food chain or in people's bodies. And companies that use these chemicals are now required to find alternatives, or at least begin researching them. The most troublesome chemicals must be registered first, but 32,000 are expected to be registered within eleven years.

Though critics of REACH complained that the new regulations would be too expensive, the European Union estimated that it would cost companies 2.3 billion euros over eleven years—about 0.04 percent of annual sales. Plus, public health analysts argued, these costs would be recouped many times over in reduced illnesses from chemical exposure—preventing 4,500 occupational cancer cases each year, for starters, and $69 billion in medical outlays over three decades. And this doesn't even include the global markets created for nontoxic alternatives.

To give you an idea of how threatening the REACH legislation was to American chemical companies, consider this: American chemical companies, which once ran trade surpluses in Europe, now run a $28 billion deficit, a number that may well increase once American consumers can buy less toxic European goods. Europe's GNP passed that of the United States in 2005, and the gap is growing. Europe is now the most important trading partner for every continent except Australia, and with this regulatory vision is setting standards once set by United States. At times, the diverging courses have seemed pronounced. In Europe, a crash in the market for chemicals like the phthalate DEHP convinced a major German producer, BASF, to cease production in 2005. At the same time, in the United States, ExxonMobil, the other major producer, said it was confident in "the safety of phthalates in their current applications" and would not stop making it.

Far from waiting on the sidelines to see what Europe would do with REACH, the American chemical industry, and its supporters in the Bush administration, fought hard to kill it. The tag-teaming was not surprising; a congressional report noted that the industry had given out more than $21 million in campaign contributions since the start of the 2000 election cycle, with 79 percent going to Republicans. President Bush was the top recipient. Greg Lebedev, the former chairman of the American Chemistry Council, explained how the lobbying worked in a 2004 speech. "We arranged for multiple elements of our government—the Department of Commerce, the U.S. Trade Representative, the Environmental Protection Agency, and the Department of State—all to express the understandable reservations about this proposed rule and its trans-Atlantic implications. I only wish that we could exert so much influence every day."

A Commerce Department memo warned that "hundreds of thousands of Americans could be thrown out of their jobs" if similar legislation were ever approved here. Secretary of State Colin Powell sent out a seven-page cable to U.S. embassies around the world, saying that REACH "could present obstacles to trade" and cost American chemical companies tens of billions in lost exports.

The Bush administration dispatched emissaries to the newest E.U. members—Hungary, Poland, Estonia, and the Czech Republic, all of which were former Communist countries with much lower environmental standards than the rest of Europe—asking them not to support the plan. Robert Donkers, an E.U. official who helped oversee Europe's chemical regulatory laws, came to the United States in 2003 to explain REACH, and was astonished at the aggressiveness of the American arm-twisting. Imagine, he said, if European companies tried as hard to intervene in American politics. "It wouldn't be tolerated," he said. "We wouldn't last ten minutes."

American public health groups, needless to say, were allowed no such influence on the Bush administration's lobbying efforts. On November 11, 2002, more than fifty public health workers, labor unions, children's health advocates, environmental organizations, and community groups wrote to Bush, urging the administration to stop undercutting European chemical reform. The administration's lobbying "runs counter to the public interest and to the transparency that is critical to our democracy," the letter said. Nearly a year later, another coalition of groups wrote Bush, requesting "that you instruct key officials within your administration to stop using federal funds to undermine this important proposed legislation, and seek ways to support progressive reform of chemicals policy that benefits public health."

Neither group ever got a response.

Nonetheless, REACH was approved on December 13, 2006. Overnight, Europe strengthened the oversight of 32,000 chemicals in a $11 trillion market affecting 500 million consumers. And it started a wave that began moving across the Atlantic. It didn't exactly crash in on Washington, D.C. But it did hit the coast of Maine.

When she first heard about REACH, Lauralee Raymond remembers, she thought, "Oh, my God! In the United States we are getting the leftovers. They are making two kinds of rubber ducks, one for Europe without phthalates, and one with phthalates, for us. We're not a Third World country. How did this happen?"

Mike Belliveau, who directed the body burden study, saw the REACH legislation as an instructive lesson. Since American federal legislation has been so strongly controlled by the chemical industry, he figured, state and local health advocates would have to try new tactics. And nothing, it turns out, works better than getting people to pay attention to the chemicals contaminating their bodies.

"I've been doing environmental advocacy work for thirty years, both in California and here in New England, and this report got the greatest media response I've ever seen," Belliveau said. "There has been a conscious shift over the last six or eight years to reframe this issue as a family health issue. If you poll on biodiversity, you get two to ten percent of people who say they are concerned. If you poll on chemicals found in food and water, you get eighty to ninety percent."

He added, "If the first reaction of the public of a policy maker is 'Oh, this is just another environmental issue,' we'll get marginalized. This is a much more populist approach. When mercury became an issue up here, we got public health people to appear before the environmental affairs committee. It wasn't just about loons and fish, it was about newborn babies and brain damage."

In other words, it's all about getting women of childbearing age to start paying attention to toxic chemicals. Especially when one of these women already happens to know about her own toxic body burden—and happens to be the state speaker of the house.

The hallways outside Hannah Pingree's office are lined with photographs of her predecessors going back 150 years. All but one are pictures of men, almost all of them bearded. To my eye, the beards, and the men wearing them, fell into three categories: Henry David Thoreau, Walt Whitman, or John Brown. There was only one exception, Libby Mitchell, who was the first woman to hold the post, ten years ago. Hannah Pingree is the second. Suffice it to say that banning toxic chemicals was not on the agenda of most of her predecessors.

"The more I worked on these issues, the more outraged I got," Pingree told me. "I want to have kids, I want to change my lifestyle, but the reality is, you can't change your lifestyle enough to get rid of all this stuff. Here we have Europe banning this stuff, and in this country, the chemical industry is lobbying Congress to increase 'safety standards' in products like mattresses and electronics. It's almost impossible for a very conscientious person to deal with."

After successfully pushing through a ban on penta and octa a few years before, Pingree decided to try for an even bigger fish: deca, a flame retardant that the chemical industry has gone all out to defend. When California debated a similar bill, the chemical industry reportedly spent $10 million to kill it, and the bill failed by two votes.

But California is one place, and Maine is another. As word got around that Maine was considering the ban, dour-looking men in suits came from outside the state to attend committee hearings. The brominated flame retardant industry hired the giant public relations firm Burson-Marsteller and set to work convincing the people of Maine that its chemicals were necessary to save lives. "We can't take a chance on fire safety," one advertisement urged. In a statement, one manufacturer, the Chemtura Corporation, said that hundreds of studies have "concluded that DecaBDE was safe for continued use."

"They ran twenty-seven full-page newspaper ads, they ran radio and television ads. This was an unprecedented campaign in Maine's history," Mike Belliveau told me. "And this was all before the bill was even voted on in committee. They were telling people that babies were going to be burning up in buildings."

This, it turns out, did not sit well with certain Maine legislators.

"Mainers don't like outsiders coming in and telling them what to do," Pingree said. "This entire brominated flame retardant army started showing up. They produce this stuff by the barrel, and here they are coming and telling us why all this stuff is safe, that all these studies were false. Maine was very frustrating for them, and it's very frustrating for outside lobbyists in general. We have very small districts, very clean elections. No one is being paid a lot to be here. As a legislator, you do not want to be seen voting for the chemical industry over a lot of young mothers.

"It was amazing, watching how much effort they put into defeating us," Pingree added. "The brominated industry must have spent at least a million bucks on television advertising, telling people to call their legislators and vote against it. They had full-page ads in all the papers. They hired all these lobbyists. Even so, we passed it almost unanimously in the house and senate. It was one of the biggest PR campaigns I've seen run, and it had almost no impact.

"They really overstepped in Maine," she remarked. "We had the firefighters on our side, and because of the industry's ad campaign, everyone in the state knew about it. There were something like two thousand bills before the legislature that year, and all these ads brought our bill right up to the top."

When the day came for a vote, supporters of the bill filled the chamber: pregnant women, farmers, doctors. The International Association of Fire Fighters was on board, Pingree said, because when PBDEs burn, they are highly toxic, and firemen can face acute exposure. On the other side were five people: four who worked in the chemical flame retardant industry and one man who, Pingree said, was "a burn victim who had been paid by the chemical industry to come to Maine and testify against our bill."

In May 2007, Pingree's bill, An Act to Protect Pregnant Women and Children from Toxic Chemicals Released into the Home, made Maine the second state in the country to ban deca, passing the law just a few months after a similar measure was approved in Washington State. The bill, which passed with overwhelming bipartisan support, banned the use of the toxic flame retardant deca in mattresses and upholstered furniture as of January 1, 2008, and phased out its use in televisions and other plastic-cased electronics by January 1, 2010. Suddenly, Hannah Pingree—"Toxic Hannah," as she had become affectionately known—was the face of a new generation fighting toxic chemicals. After CBS News came to town to do a story on her bill, Pingree started getting calls from people all over the country; they'd ask her questions like what kind of car seat to buy for their children.

But she wasn't done. In late February 2008, when Maine's state legislators filed past the state capitol's Hall of Flags, they were greeted by a giant rubber duck. This was not your ordinary rubber duck. Its head nearly touched the room's ceiling, which is thirty feet high.

Tired of going after individual chemicals, Hannah Pingree had decided to follow the lead of Europe's REACH legislation and pursue a broader, more comprehensive toxics policy. The bill would go far beyond the banning of individual chemicals and, instead, establish a database of several thousand chemicals of "high concern." It would then create a list of "high-priority" chemicals that would merit immediate regulation. Importantly, the law followed the European standard of proving "exposure" rather than "risk": if a chemical is detected in people, or in wildlife, or is made in sufficiently high volume, businesses are required to disclose its presence in their products, what the purpose of the chemical is, what the chances are that the chemical will be released, and whether there are safer alternatives available "at a comparable cost." If reasonable alternatives are available, the state can ban the products that contain it.

The bill was supported by the American Cancer Society, the American Lung Association, and the Maine Chapter of the American Academy of Pediatrics. It was opposed by the American Chemistry Council. "States are ill-equipped, under-funded, and do not have the skill set, at this time, to take on this complicated and difficult endeavor," a spokeswoman for the industry group said.

During the hearing, Matt Prindiville, a policy analyst with the Natural Resources Council of Maine, told legislators that recent news stories about lead-contaminated toys arriving from China offered the barest glimmer of how many toxins consumers—and their children—are exposed to every day.

"Most of you may know that it's already illegal to paint a toy with lead paint, but did you know that it's perfectly legal to put lead into plastic toys as a stabilizer?" Prindiville said. "When we

participated in toy testing, we found that the highest lead concentrations were not in painted wooden toys like Thomas the Tank Engine. They were in vinyl backpacks and lunch boxes and soft vinyl toys marketed to toddlers. And that's just lead. We've known lead's bad for kids for a long time. We took it out of gas, paint, and other products, but it's still coming back to haunt our children. The scarier fact is that 90 percent of the chemicals in products marketed to our kids have never been adequately tested for health and safety. What's worse is that we know some of them are linked to learning disabilities, cancer, and other health risks and that we're finding them in our bodies and in our children's bodies."

Prindiville held up a pair of rubber ducks: one made of PVC plastic containing phthalates, the other made of natural latex. "Which one do you want your toddler sucking on?" he asked. "My daughter was born in June of last year. She's nine months old and loves to put things in her mouth. She chews on anything and everything. She's like a puppy. If I were to go to the store to buy a rubber ducky for my kid, there is no way of knowing which one to buy. This one doesn't say, 'WARNING: I have hormone-disrupting phthalates in me.' This one doesn't say, 'I don't have any toxic chemicals. I'm perfectly safe to bring home.' Sure, you can find out the unit price. You can find out where it's made. It'll tell you that it comes in red, yellow, and green. It'll tell you that it's made out of PVC, but it won't tell you if it has phthalates in it, or bisphenol A or lead.

"All we're saying is that Maine people have the right to safe products for their kids and Maine DEP has the responsibility to identify those products that are out there that have safer alternatives to toxic chemicals, like this rubber duck. My nine-month-old daughter and all the other kids in Maine today, all the pregnant women and families, the only way for us to get kid-safe products is to flock together and call on our elected officials, and say, 'You can't duck this issue anymore.'"

Bad puns aside, Prindiville and his allies carried the day. Suddenly Maine not only had the country's most comprehensive laws on toxic chemicals, it had pushed the entire conversation on toxics in a new direction. Not only had a state acted more stringently than the federal government, it had adopted a European-style policy that gives the benefit of the doubt to public health rather than the chemical industry. It had pushed the burden of proving a chemical's safety onto the shoulders of manufacturers.

"What's important is that the safety system has been badly broken," Mike Belliveau told me. "Industry is really good at defending individual products. If you attack deca, they'll show you loads of studies saying it's not bad for you. If you attack phthalates, they'll roll out all their experts, the whole product-defense campaign. But when you're not proposing to ban BPA or phthalates, but trying to overhaul the whole system, the industry is at a loss. The individual chemical people—the bromated people, the BPA people—they're at a loss. All they can say is 'The federal law is okay as it is.' And that law's a joke. TSCA hasn't been updated in thirty years. It's completely absurd."

The Pingree family, at least, is nudging the federal government to follow Maine's example. Hannah's mother, Chellie, Maine's first-term congresswoman, recently introduced legislation to ban deca nationally. Soon after she called a press conference to announce the bill, the chemical industry announced that it would voluntarily stop producing deca within three years. Pingree vowed to push ahead with her bill, to make sure that the industry "doesn't start using another chemical that is just as dangerous."

There are other positive signs at the federal level. A couple of years ago, researchers at the National Institute of Environmental Health Sciences began revamping the federal government's National Toxicology Program, which sets standards for how chemicals are tested. Over about seven years, they hope to develop a series of lab tests that will ultimately screen some 100,000 industrial compounds, individually and in mixtures, for biochemical markers such as effects on specific genes. The chemicals will then be ranked by mechanism of action and suspected toxicity and assigned priorities for further study. "It's taken us twenty-five years and $2 billion to study 900 chemicals," Dr. Christopher Portier told *The Wall Street Journal*. "If this works, we can study 15,000 in a year."

None of this can come too soon. If we had acted on what we knew to be true about the dangers of industrial chemicals, we could have prevented a million and a half deaths, writes Devra Davis, the director of the Center for Environmental Oncology at the University of Pittsburgh Cancer Institute. "There is no one who deals with the disease now who doubts that we need to open a new front. To reduce the burden of cancer today, we must prevent it from arising in the first place. No matter how efficient we become at treating cancer, we have to tackle those things that cause the disease to occur or recur."

To heal our poisoned bodies and to alter our poisoned landscape, we must first of all begin to see things more clearly. We must become more mindful. Spraying synthetic chemicals on our lawns, in this way of thinking, is mindless; planting native species is mindful. The same approach applies to choosing a mattress, toys for our children, a glass of water, a bag of grapes. Once we have access to information, once we begin paying attention, we develop a fuller, more personal connection to the actions we take. Once we have information, we have no choice but to choose between continuing to delude ourselves and changing the way we live our lives. We can't unlearn what we have learned.

There is an important difference between innocence and naïveté: the first implies a lack of knowledge, the second a lack of responsibility. Thought of on a grander scale, this is the difference between looking for cures to problems and looking to prevent problems in the first place. It's not as if this has to be so complicated, after all; people managed to live pretty well before synthetic chemicals were introduced, and no doubt we will continue to live pretty well long after they are gone. The question is how long this transition will take.

Acknowledgments

The completion of this book owes a deep debt to a wide range of people. In Maine, thanks to Carrol Lange and Mike Belliveau of the Environmental Health Strategy Center and to Matt Prindiville of the Natural Resources Council of Maine. Special thanks to Russell Libby, Lauralee Raymond, Amy Graham, Hannah Pingree, Rick Churchill, Tim Feeley, and Paul Tukey.

In Baltimore, thanks to Albert Donnay for his expertise and his friendship. Thanks also to Angela Smith of the Center for a Livable Future at Johns Hopkins, and Tom Thompson and Gerry Hanlon and Gino Freeman. Thanks to Ken Oldham and Raja Veeramachaneni at the Maryland State Highway Administration, for showing how roadsides can be maintained without the excessive use of synthetic chemicals.

At the University of Delaware, thanks to Hal White, Al Matlack, Belinda Orzada, Matt Kinservik, Ben Yagoda, and Dawn Fallik. Special gratitude to Doug Tallamy and Jerry Kauffman for their expertise, cooperation, and collegiality.

Thanks to all of my students in The Literature of the Land, but especially to my research assistants Maddie Thomas, Wallace McKelvey, and Sean Mis.

In New York, thanks to Wendy Gordon and Wes Davis.

In Vermont, thanks to Martin Wolf, and in Oregon, thanks to Wendy Buchanan.

My debt to the writers and scientists devoted to the issues discussed here is evident throughout these pages, but I would like to single out for special admiration the work of Devra Davis, Sandra Steingraber, Philip Landrigan, Ellen Silbergeld, Theo Colburn, Dianne Dumanoski and John Peterson Myers, Stacey Malkan, William McDonough and Michael Braungart, and Ted Steinberg.

At Random House, thanks to Jennifer Hershey, Caroline Sutton, Laura Ford, Courtney Moran, and, especially, Andy Ward, who came in at just the right time and helped steer this project through the weeds. Bonnie Thompson was a razor-sharp copy editor. Neil Olson continues to be the apotheosis of agents.

Finally, thanks to Katherine, Steedman, and Annalisa, who nourish me in ways too numerous to list here.

Appendix

When it comes to general rules about ways to reduce our exposure to synthetic chemicals, it's hard to do better than this: use products and techniques your great-grandmother used—provided she was born before World War II, when most of the chemicals in use today got their start. "We clean for health," writes Ellen Sandbeck, the author of *Green Housekeeping*. "If a woman living in 1904 was transported one hundred years into the future and was given a washing machine and dryer, a vacuum cleaner, a sewing machine, a dishwasher, running water, indoor plumbing, electric lights, and a gas or electric stove, would she spend all that saved time fretting over bacteria in her drain or garbage pail? Our great-grandmothers were obviously clean enough, or we wouldn't exist. There is not a synthetic cleaning product on the market that can improve our health, though there are plenty that can ruin it." The same philosophy, of course, can be applied to our other activities.

Sprinkled throughout this book are dozens of simple suggestions for reducing your exposure to toxic chemicals. All will benefit your family's health. Most will save you money. Some may even connect you more deeply to your home, your garden, and your kids. Here, in short form, are a few more ideas.

GENERAL RESOURCES

A number of environmental and health organizations conduct research and work to change legislation regulating toxic chemicals in consumer products. Here are some of them:

The Center for Science in the Public Interest (cspinet.org) is a good resource for information on food safety and health science.

Clean Production Action (www.safer-products.org) is a useful resource for green chemistry and alternative materials.

Environmental Working Group (ewg.org) specializes in research and the creation of databases on body burdens and toxics in consumer goods from beauty products to agriculture and drinking water.

The Environmental Defense Fund (edf.org), one of the country's leading environmental groups, uses scientific research to support both public interest lawsuits and political lobbying on behalf of sound environmental and health policy.

Greenpeace, a global environmental organization, has taken a leading role in the regulation of toxic chemicals. Its analysis of persistent organic pollutants, or POPS, can be found at www.greenpeace.org/usa/campaigns/toxics/toxics-reports/pops.

The Natural Resources Defense Council (nrdc.org), one of the country's largest environmental advocacy groups, uses lawsuits and political lobbying to push for better environmental policy.

The Silent Spring Institute (coalition.silentspring.org) is a collation of scientists, physicians, and public health advocates working to understand the connection between environmental contamination and women's health.

The Union of Concerned Scientists (ucsusa.org) advocates for the use of science in government policy overseeing everything from the spread of toxic chemicals to issues of climate change and renewable energy.

The United States Public Interest Research Group (uspirg.org). Along with its many state affiliates, U.S. PIRG lobbies for government reform on everything from product safety to toxins in the environment.

Some good resources for a list of nontoxic consumer products are the *Consumer Reports* Greener Choices website (greener-choices.org) and the Organic Consumers Association (organic-consumers.org). Greenseal (greenseal.org) tests all kinds of consumer products for their effects on health and the environment.

To learn more about your state's legislative efforts to curb the spread of toxic chemicals in consumer products, check out a database compiled by the University of Massachusetts–Lowell's Center for Sustainable Production: chemicalspolicy.org/uslegislationsearch.php.

The website for the federal Consumer Product Safety Commission, which is charged with overseeing the safety of a vast array of products, can be found at cpsc.gov/.

BODY BURDEN STUDIES

Health advocacy groups have conducted body burden studies in a number of states. Here are a few:

Washington: Pollution in People (pollutioninpeople.org/)
Oregon: Oregon Environmental Council: (www.oeconline.org/our-work/kidshealth/pollutioninpeople)
Maine: Alliance for a Clean and Healthy Maine (www.cleanandhealthyme.org/)
The Environmental Working Group's study "Pollution in Newborns": www.ewg.org/reports/bodyburden2/
The Centers for Disease Control and Prevention's "National Report

on Human Exposure to Environmental Chemicals": www
.cdc.gov/exposurereport/

HOME

When it comes to "freshening" the air in your house, try doing it
the old-fashioned way: open your windows. Air fresheners typically contain fragrances made with phthalates as well as neurotoxins, and the tiny aerosolized droplets can easily be absorbed into
the skin and the lungs. A good rule of thumb: don't try to disguise
smells. For similar reasons, avoid synthetic potpourri. Instead, try
boiling cinnamon and cloves in a pan of water. Or try placing cedar
blocks and sachets of dried flowers around the house. Also try po-
sitioning pots of fragrant houseplants in your kitchen—they add a
nice aroma and absorb airborne toxins, to boot. When cleaning up
kitchens and bathrooms, use vinegar and baking soda.

Avoid aerosolized indoor pesticides. Instead of "bombing" your
house for fleas, comb and bathe your pets regularly. A vacuum is
an effective weapon in the fight against fleas; before vacuuming,
try sprinkling borax on infested areas. If need be, spray fleas on
your pet with citrus oil or concentrated lemon water.

Avoid synthetic carpet-cleaning chemicals. Vacuum twice
weekly, preferably with a vacuum cleaner equipped with a HEPA
(high-efficiency particulate air) filter.

Use high-quality furnace filters. Furnace filters are rated according to their minimum efficiency rating value, or MERV; the
higher the MERV rating, the better. Have your air ducts inspected,
and cleaned if necessary.

Always remove your shoes when you walk in the front door;
shoes track in an astonishing variety of toxins and bacteria.

CLOTHING

Clothes labeled as "permanent press" may contain formaldehyde;
other synthetic fabrics made be treated with toxic stain-resistant
or flame retardant chemicals. When shopping for clothes, look for

items made from natural (and preferably organic) materials like cotton and wool. Wash new clothes before wearing them to reduce exposure to formaldehyde, often added to make clothing appear unwrinkled.

For children's pajamas, avoid synthetic materials treated with flame retardants. Dress kids in cotton instead.

When getting your clothes dry-cleaned, seek out organic cleaners who do not use the chemical perc.

THE BATHROOM

Another good general rule: Avoid using products made with ingredients you can't pronounce. This is as true for cosmetics and shampoos as it is for food.

Be careful when shopping for personal care products, especially when they are for your babies or small children. Baby wipes, deodorants, shampoos, fragrances, hair gels, hairsprays, and hand lotions can all contain hormone-disrupting phthalates. Cosmetics and personal care products contain a vast array of largely unregulated synthetic chemicals, from formaldehyde in baby soaps to lead in hair dyes to coal tar in dandruff shampoos and skin creams. Horst Rechelbacher, who built Aveda into the largest nontoxic beauty salon chain in the United States before selling it to Estée Lauder in 1997, had it right: "If you wouldn't put it *in* your body, why would you put it *on* your body?"

Since many beauty products are less than clear about their ingredients, look for products to buy or avoid on the Environmental Working Group's Skin Deep cosmetics database: www .cosmeticsdatabase.com/. The Campaign for Safe Cosmetics pushes for legislation and regulation of personal care products. Its database (www.safecosmetics.org) is a good resource if you want to learn more about both laws and responsible companies.

Conventional toothpaste may contain toxic ingredients like ethylene glycol, which is also used in paint and antifreeze. Tom's of Maine makes plant-based toothpaste and other personal care products (tomsofmaine.com).

Synthetic drain cleaners are among the most toxic of all household chemicals. Instead, try using a half cup of baking soda followed by a half cup of white vinegar. Better yet: use a plunger.

THE LAUNDRY ROOM

Not only are conventional detergents made with petroleum products and phthalate-laden "fragrances," but they can contain phosphates, which are harmful for local waterways. Instead, try plant-based detergents. Other nontoxic substances, such as washing soda (sodium bicarbonate), are good stain removers for clothes as well as for stains on other types of fabrics. Instead of chlorine bleach, try using an equal part of borax or washing soda for every part of laundry detergent you add to the wash. For stains, use borax, lemon juice, hydrogen peroxide, or white vinegar. If you prefer store-bought products, look for nonchlorine bleach.

Fabric softeners can contain neurotoxins like isopropylbenzene and possible carcinogens like styrene; respiratory irritants like xylene and phenol; and phthalates. Instead, add a quarter cup of white vinegar or baking soda to the wash. For commercial products, look for vegetable-based (frequently coconut oil) surfactants and softeners.

Avoid using dryer sheets, which contain a variety of toxic chemicals. Instead, hang your laundry to dry, either outside or inside. If you live in a community that prohibits the hanging of laundry, work to change the rules. See laundrylist.org

THE KITCHEN

When buying cleaning products, watch for such words on the label as "danger," "poison," "harmful vapors," or "skin irritant." Also be skeptical of vague terms like "eco-friendly" or "green," which can mean almost anything. Look instead for specific terms like "solvent-free," "plant-based," "no phosphates," or "no petroleum ingredients." Less toxic cleaning supplies are available from a number of companies, including Burt's Bees (burtsbees.com), Dr. Bronner's (drbron ner.com), Ecover (ecover.com), and Seventh Generation (seventh generation.com). Supplies for making cleaning products can be found at fromnaturewithlove.com and snowdriftfarm.com.

Avoid using nonstick cookware, which is often made with Teflon chemicals. Instead, use stainless steel or cast iron and sauté food in butter or olive oil.

Use the ventilation hood on your stove; be sure the hood actually works and vents to the outside of the house. Outfit your kitchen with a working carbon monoxide monitor, and make sure it works, too. Check that your furnace is properly ventilated to the outside of the house. Likewise your garage. Avoid pumping car exhaust into unventilated garages, especially in winter.

Use dish soaps that are clear, since colors can contain dyes contaminated with lead and arsenic. As with other cleaning agents, try to use soaps made with plant oils. Castile soap (like that made by Dr. Bronner's) are effective. Use dishwasher soaps that contain borax.

Instead of spending money on petrochemical cleaning products or toxics like chlorine, try making your own household cleaners using nontoxic (but effective) ingredients like baking soda, white vinegar, hydrogen peroxide, and lemon. A bucketful of hot water with a cup of white vinegar and a drop of dish liquid will go a long way to getting your floors clean.

Seek out companies that make plastics out of natural (and nontoxic and biodegradable) ingredients like corn or potatoes. Nat-Ur, for example, makes everything from biodegradable dinnerware to compostable garbage bags. See Nat-urstore.com.

When buying napkins, paper towels, coffee filters, toilet paper, and, especially, tampons, seek out products made without chlorine bleach. Hint: if it's brown, it's likely been made without chlorine. Use washable rags to clean up instead of paper products.

FOOD

Whenever possible, shop for food grown without pesticides.

Avoid processed foods and drinks made with dyes, which may contain coal tar.

Use glass baby bottles instead of plastic, especially for warm liquids, which can cause toxic chemicals like phthalates and bisphenol A to leach from the bottle into the liquid. Use clear (rather than amber-dyed) nipples, which are made of silicone and do not contain bisphenol A.

Do not buy food in a big box store unless it's organic. Do not eat microwave popcorn packaged in grease-resistant packaging. Buy popcorn in glass jars instead.

Whenever possible, buy organic meat raised without hormones or antibiotics. Avoid meat that can be kept in the refrigerator for a long time, since it may contain the toxic preservative sodium nitrate.

When buying beef or milk, try to find products from animals raised on grass, their natural food source, rather than corn.

Avoid buying food packaged in cans, virtually all of which are lined with plastic, which can contain bisphenol A. Instead, wherever possible, buy food packaged in glass.

Eat wild fish, not farm-raised fish, which can be tainted with, among other chemicals, toxic flame retardants.

Avoid wrapping food in plastic cling wrap, which is often made of PVC and may contain bisphenol A.

TOYS

When shopping for toys, look for products made from wood. Avoid plastics (which can contain phthalates), especially for teething infants and young children. A database of healthy toys can be found at healthystuff.org. Another organization working to change laws governing toys and other products, MomsRising, can be found at www.momsrising.org.

Buy toys from companies that boast about the ingredients in their toys—and back it up. Buy only watercolor paints, which typically release only low levels of (or no) volatile organic compounds.

Avoid serving your kids food in lunch boxes with the following recycling codes: 3 refers to PVC, which can contain phthalates;

6 means polystyrene, a possible human carcinogen; and 7 usually indicates polycarbonate, which can contain bisphenol A. A good resource for nontoxic, reusable lunch boxes and bags is www .reusablebags.com/.

FURNITURE AND APPLIANCES

When shopping for computer equipment (which can contain everything from mercury to flame retardants), consider checking Green Electronics Council's Electronic Product Environmental Assessment Tool (EPEAT): www.epeat.net/. Many computer companies, including Apple, Hewlett-Packard, IBM, and Compaq, have recycling programs. Check the company website for more. The National Recycling Coalition is a good resource for electronics recyling: nrc-recycle.org. For general information on electronic products, check the Green Electronics Council (greenelectronics council.org).

Instead of furniture made with laminated plywood or particleboard (the particles are often glued together with formaldehyde), try buying solid wood furniture, preferably made with materials certified by the Forest Stewardship Council (fscus.org). Be sure that whatever plywood you use was made without formaldehyde. Check out the Sustainable Forestry Initiative (sfiprogram.org).

Instead of synthetic (and aerosolized) wood furniture polishes, which can contain volatile organic compounds as well as toxic fragrances, try using olive oil.

Buy upholstery made of natural fabrics like cotton (preferably organic) or wool, hemp or linen. Synthetic upholstery is typically made of petrochemicals and is often treated with stain-resistant chemicals, which contain toxic perfluorochemicals. Avoid cushions made with PBDE flame retardants. Instead, buy cushions made with untreated foam or with all-wool stuffing, which is naturally flame-resistant.

Avoid mattresses made with PBDE flame retardants. Instead, look for mattresses stuffed with wool or made from natural latex.

"Memory foam" can contain toluene. For blankets, try using wool and cotton instead of polyester and acrylic. Look for pillows stuffed with feathers or wool rather than polyester or polyurethane foam.

HOME MAINTENANCE

Throw out old cans of paints and tubes of caulk and the like, which contain a Pandora's box of toxic chemicals and can release them as volatile organic compounds.

Avoid wallpaper and blinds and shower curtains made of polyvinyl chloride (PVC) plastic; instead, use curtains made of natural fibers like cotton, silk, or wool.

Avoid using synthetic polyurethane on wood floors; use water-based urethane instead. Look for low-VOC water-based paints instead of oil-based paints. When painting inside, be sure to allow for plenty of ventilation. Nontoxic paints and finishes can be found at the Real Milk Paint Company (realmilkpaint.com) and William Zinsser and Co. (zinsser.com).

A good place to look for less toxic flooring and carpeting is www.greenfloors.com.

Popcorn is meant to be eaten, not sprayed on ceilings.

Avoid wall-to-wall carpeting, which is typically made of synthetic fibers, is often laid over padding made with toxic chemicals and glued to the subfloor, and can become a sink for indoor pollutants. Instead, use area rugs made of natural fibers—cotton, wool, or hemp. For greener alternatives, check out the Carpet and Rug Institute (carpet-rug.org).

Good resources for healthy building materials can be found through the Healthy Building Network (healthybuilding.net), the Institute for Local Self-Reliance (www.ilsr.org), the Healthy House Institute (healthyhouseinstitute.com), or the U.S. Green Building Council (usgbc.org).

THE TAP

Drink municipal tap water, which is imperfect but still more regulated than commercially bottled water. Drinking from the tap eliminates the need for plastic bottles and typically costs less than twenty cents a gallon. If you get your water from a well, have it tested regularly. To learn more about your community's drinking water, check out Clean Water Action (cleanwateraction.org).

To find out how your community treats its wastewater and storm water, check out the Water Environment Research Foundation (werf.org).

Don't drink hot liquids from cups made from plastic, which can leach into the drink. Instead, use glasses made of . . . glass. Or stoneware. Even paper cups are typically bleached with chlorine and are commonly coated with plastic resin.

Buy a reusable steel water bottle. If you're in the market for a reusable hard-plastic water bottle, be sure it is not made with bisphenol A.

An end-of-tap carbon water filter can help remove chlorine by-products, pesticides, and some heavy metals, but is not effective at removing bacteria, nitrates, or arsenic.

Be conscious of the way you dispose of your pharmaceutical drugs. Avoid flushing unused doses down the toilet, since these will inevitably end up in local water supplies.

THE LAWN

Get rid of your lawn-care company or hire one that uses only organic gardening practices. To learn more about pesticides and their alternatives, visit the Pesticide Action Network North America (www.panna.org) or Beyond Pesticides (www.beyondpesticides.org). The National Pesticide Information Center at Oregon State University can be found at npic.orst.edu.

Consider replacing some of your grass with shrubs and trees native to your region. Not only will this reduce your need for synthetic chemicals, it will have an important (and visible) impact on local wildlife, especially birds. To learn more about which native species to plant in your region, check out PlantNative (www.plantnative.org), American Native Plants (www.americannativeplants.net/), or Wild Ones (www.for-wild.org/). Many states have native plant societies that offer seminars and planting advice.

For more information on organic lawn care, visit www .safelawns.org. Another good resource for lawn care and gardening is the Rodale Institute (rodale.org). To learn more about building backyard habitats for birds, consult the National Audubon Society (audubon.org).

When confronted with garden pests, consider spraying plants with water infused with soap, lemon, or hot peppers.

Avoid using glyphosate (Roundup), dicamba, 2,4-D, and the myriad other synthetic herbicides on the market. Instead, mulch your gardens to get rid of weeds or use a hoe to dig them up. Consider releasing beneficial insects (ladybugs, praying mantises, green lacewings, predatory nematodes) into your garden to combat pests like aphids and Japanese beetles. Planting native species also attracts birds, which will help keep insects in balance.

Trap rainwater in rain barrels fixed to the ends of downspouts. This will not only help relieve runoff but will offer a free supply of water for gardens during drought. Be sure to fit the barrel with a screen to avoid breeding mosquitoes. A good source for rain barrels, made from retrofitted Greek pickle barrels, is rainbarrelsand-more.com.

Consider replacing gas-powered lawn mowers with electric mowers or push (reel) mowers. You can reduce a lawn's nitrogen requirements by 50 percent just by leaving lawn clippings on the grass. Encourage the growth of clover, which doesn't need mowing and fixes nitrogen in the soil (thus eliminating the need for synthetic fertilizers).

Instead of a gas-powered leaf blower, consider using a rake. Encourage children to jump in the piles.

Notes

PROLOGUE

12   **"What most struck me"**: Alan Riding, "1,800 Objects from the Titanic: Any Claims?" *New York Times*, December 16, 1992.
    The Synthetic Century: See Michael Pollan, *The Omnivore's Dilemma* (New York: Penguin Press, 2006). See also Douglas Fischer, "The Great Experiment," the second part of a three-part series in the *Oakland Tribune*, March 10, 2005.
    By the end of the Second World War: *New York Times*, May 16, 1940.

13   **The trouble with:** U.S. PIRG Education Fund, "Toxic Pollution and Health: An Analysis of Toxic Chemicals Released in Communities Across the United States," March 2007, p. 4.
    And since these chemicals: "I have no doubt that some of these things that we're discovering out there have been there since the dawn of the plastic era in the 1950s," Moore told CBS News. For sea animals—birds, fish, seals—"it's like putting them on a plastic diet. It becomes part of their tissue." By some estimates, plastic kills a million birds and 100,000 marine mammals a year. Rob Krebs, a spokesman for the American Plastics Council, counters that just because plastic is everywhere, "it shouldn't be the whipping boy of environmental-

ists. It's good material, and so when we talk about Charles Moore, we really need to look in the mirror. We need to look at ourselves." "Sailing the Seas of Trash: Vast Area of Pacific Ocean Polluted with Plastic," CBS News, January 6, 2004; KQED, August 22, 2008. See also Alan Weisman, "Polymers Are Forever, *Orion*, May–June, 2007.

14   **I was born:** Rachel Carson, *Silent Spring* (Boston: Houghton Mifflin, 1962), pp. 28–30.

Carson's book was savaged: "Rachel Carson Dies of Cancer; 'Silent Spring' Author Was 56," *New York Times*, April 15, 1964.

15   **In addition to pesticides:** Anita Lahey, "Unsafe Assumption," *Canadian Geographic*, May–June 2003.

In short, the middle decades: Michael T. Kaufman, "Waldo Semon Dies at 100; Chemist Who Made Vinyl," *New York Times*, May 28, 1999.

By 2004, the U.S. chemical industry: Nena Baker, *The Body Toxic: How the Hazardous Chemistry of Everyday Things Threatens Our Health and Well-Being* (New York: North Point Press, 2008), pp. 39–40. See also Katharine Mieszkowski, "Plastic Bags Are Killing Us," *Salon*, August 10, 2007; Elizabeth Royte, "A Fountain on Every Corner," *New York Times*, May 23, 2008; Athanasios Valavanidis et al., "Persistent Free Radicals, Heavy Metals and PAHs Generated in Particulate Soot Emissions and Residue Ash from Controlled Combustion of Common Types of Plastic," *Journal of Hazardous Materials*, vol. 156, nos. 1–3 (August 15, 2008), pp. 277–84.

16   **In 1990, the National Academy of Sciences:** Anita Lahey, "Unsafe Assumption," *Canadian Geographic*, May–June 2003. See also Yana Kucker and Meghan Purvis, "Body of Evidence: New Science in the Debate over Toxic Flame Retardants and Our Health," National Association of State PIRGs and Environment California Research and Policy Center, 2004.

Every day, the United States: Michael Wilson, with Daniel A. Chia and Bryan Ehlers, "Green Chemistry in California: A Framework for Leadership in Chemicals Policy and Innovation," California Policy Resource Center, University of California, 2006. You can find the report at www.ucop.edu/cprc/.

Here's the problem: Sandra Steingraber, *Living Downstream: An Ecologist Looks at Cancer and the Environment* (Reading, Mass.: Addison-Wesley, 1997), p. 92.

17   **To be sure:** Douglas Fischer, "The Great Experiment," *Oakland Tribune*, March 10, 2005.

In the spring of 2010: Nicholas Kristof, "Do Toxins Cause Autism?" *New York Times*, February 25, 2010.

Today, despite vast sums: Indeed, figuring out how to get people to live longer with cancer is not the same thing as reducing the number of people who get cancer in the first place. In 1971, President Richard Nixon—fresh off watching an American walk on the moon—signed the National Cancer Act, pledging to find a cure for cancer within five years. He did not ask the country to reconsider its exposure to tobacco, or industrial chemicals, or asbestos, or synthetic consumer products. He pushed the War on Cancer more as something the country could rally behind—as opposed, say, to the war on the ground in Vietnam. Since then, the National Cancer Institute alone has spent $105 billion on cancer research, and this is just a part of the total spent by the government, universities, pharmaceutical companies, and charitable groups. Although cancer therapies and early diagnoses have improved some cancer survival rates, the overall death rate for cancer, adjusted for the size and age of the population, has dropped by just 5 percent from 1950 to 2005. By comparison, the death rate for heart disease has dropped by 64 percent, and for flu and pneumonia by 58 percent. Some scientists and public health experts worry that the much-ballyhooed search for a cancer "cure" has led to a widespread delusion: however bleak the diagnosis, some people believe miracle drugs or the latest technology will somehow be there to save them. A Herblock cartoon, printed in the *Washington Post* in 1977, shows a harried cancer researcher bent over a microscope, flanked by a fearsome gang of industry representatives from Big Tobacco, Big Chemical, and Big Asbestos. Meanwhile, a government official begs the scientist, "Could you hurry and find a cure for cancer? That would be so much easier than prevention." See Richard Horton, "Cancer: Malignant Maneuvers," a review of Devra Davis's *The Secret History of the War on Cancer*, in the *New York Review of Books*, March 6, 2008. See also Gina Kolata, "In Long Drive to Cure Cancer, Advances Have Been Elusive," *New York Times*, April 24, 2009. The Herblock cartoon is reprinted in Davis's book on p. 198.

And while cancer seems: Yana Kucker and Meghan Purvis, "Body of

Evidence: New Science in the Debate over Toxic Flame Retardants and Our Health," National Association of State PIRGs and Environment California Research and Policy Center, 2004. See also David Kohn, "Body Wars," *Baltimore Sun*, June 5, 2008; Felicity Barringer, "Exposed to Solvent, Worker Faces Hurdles," *New York Times*, January 24, 2009; Evelyn Hess, "Environmental Chemicals and Autoimmune Disease: Cause and Effect," *Toxicology*, vols. 181–82 (2002).

18 **The deeper one looks:** Environmental Defense Fund, *Toxic Ignorance: The Continuing Absence of Basic Health Testing for Top-Selling Chemicals in the United States.* For the full report, see www.environmentaldefense.org/documents/243_toxicignorance .pdf.

19 **Those "ingredients" lists:** See www.dtsc.ca.gov/hazardouswaste/ perchlorate/.

20 **Pick up a can:** "EPA Probes Suspected Link Between Household Chemical, Feline Hyperthyroidism," *DVM Newsmagazine*, vol. 38, no. 10 (October 2007). See also "Cats and Dogs Harbor Higher Rates of Certain Chemicals Than Their People Do," National Public Radio, *All Things Considered*, April 17, 2008.
"Information," such as it is: Mark Schapiro, "Toxic Inaction: Why Poisonous, Unregulated Chemicals End Up in Our Blood," *Harper's*, October 2007.

21 **"The effects don't just accumulate":** Along with the dark trends, of course, are astonishing gains in medical research and in cancer survival rates. In 1978, Richard Horton notes, there were 3 million cancer survivors. By 2005, there were 10 million. Breast cancer rates began to decline in 2003, perhaps because of the mass abandonment of hormone-replacement therapy. And new kinds of treatments for diseases, like monoclonal antibodies for colorectal cancer, are prolonging lives. There is a lot of money involved in these issues, Horton reminds us. Oncology has become very big business. The chemical industry, after all, is not the only powerful force in Washington. The pharmaceutical industry also leans on Congress, which then leans on the FDA to minimize government monitoring of the potentially harmful ingredients in everything from body lotions to pharmaceutical drugs and dietary supplements. Ritalin, for example, a drug taken by one in ten kids at some point in their lives, may cause liver

tumors and genetic damage, Horton notes. Yet the FDA has taken no action, and the drug remains ubiquitous. See Devra Davis, "A Thousand Threats: Cancer-Causing Chemicals Don't Work Alone, but in Tandem," *Newsweek International*, March 5, 2007. See also Sandra Steingraber, *Living Downstream*, p. 32, and Richard Horton, "Cancer: Malignant Maneuvers," *New York Review of Books*, March 6, 2008, p. 25. The figures for cancer deaths are from the American Cancer Society, *Cancer Facts and Figures 2006*, pp. 1, 22. A pdf of the report can be found at www.cancer.org/downloads/ STT/CAFF2006PWSecured.pdf.

21 **More questions: What role:** See www.epa.gov/oppbead1/ pestsales/01pestsales/usage2001.htm.

22 **More recently, experts say:** Sandra Steingraber, *Living Downstream*, pp. 20–21, 28. See also Ruth A. Lowengart et al., "Childhood Leukemia and Parents' Occupational and Home Exposures," *Journal of the National Cancer Institute*, vol. 79, no. 1 (July 1987), and Martin Belson et al., "Risk Factors for Acute Leukemia in Children: A Review," *Environmental Health Perspectives*, vol. 115, no. 1 (January 2007).

CHAPTER ONE: THE BODY

27 **To make a point:** Daniel Martineau et al., "Cancer in Wildlife: a Case Study; Beluga from the St. Lawrence Estuary, Quebec, Canada," *Environmental Health Perspectives*, vol. 110, no. 3 (March 2002). See also Robert Quiroz et al., "Analysis of PCB Levels in Snow from the Aconcagua Mountain (Southern Andes) Using the Stir Bar Sorptive Extraction," *Environmental Chemistry Letters*, vol. 7, no. 3 (September 2009); Cynthia A. DeWit et al., "Levels and Trends of Brominated Flame Retardants in the Arctic," *Chemosphere*, vol. 64, no. 2 (June 2006), pp. 209–33; Michael G. Ikonomou, Sierra Rayne, and Richard F. Addison, "Exponential Increases of the Brominated Flame Retardants, Polybrominated Diphenyl Ethers, in the Canadian Arctic from 1981 to 2000," *Environmental Science and Technology*, vol. 36, no. 9 (2002), pp. 1886–92, and P. O. Darnerud et al., "Polybrominated Diphenyl Ethers: Occurrence, Dietary Exposure, and Toxicology," *Environmental Health Perspectives*, vol. 109, supp. 1 (2001), pp. 49–68.

28 **"Our experience with"**: M. LaGuardia, R. C. Hale, and E. Harvey, "Are Waste Water Treatment Plants Sources for Polybrominated Diphenyl Ethers?" Paper presented at the Annual Meeting of the Society of Environmental Toxicology and Chemistry (SETAC), 2003.

29 **If lab science:** See www.pops.int/documents/guidance/beg _guide.pdf.

34 **The Maine study was modest:** "Body of Evidence: A Study of Pollution in Maine People," Alliance for a Clean and Healthy Maine, June 2007; available at www.cleanandhealthyme.org.

35 **The truth is:** Richard Horton, "Cancer: Malignant Maneuvers," *New York Review of Books*, March 6, 2008.

  Given the immense: Centers for Disease Control and Prevention, "Executive Summary," *National Report on Human Exposure to Environmental Chemicals*, 2005. The full report can be found at www.cdc.gov/exposurereport/.

36 **The Maine study was not:** To see the full multistate report, visit www.isitinus.org/documents/Is%20It%20In%20Us%20Report .pdf. Another useful site is www.chemicalbodyburden.org/. The Oregon report can be found at www.oeconline.org/our-work/ kidshealth/pollutioninpeople. For the full Environmental Working Group "Pollution in Newborns" report, see archive.ewg.org/ reports/bodyburden2/.

  The CDC began: The CDC study did little to interpret the data and came full of disclaimers. For one thing, even for chemicals suspected of being troublesome, it noted, "very limited scientific information is available on potential human health effects." For another, the study was not designed to estimate exposure levels by city, or even by state. For yet another, being contaminated with toxic chemicals does not necessarily mean someone will become sick. "Just because people have environmental chemicals in their blood or urine does not mean that the chemical causes disease," the CDC reported. "The toxicity of a chemical is related to its dose or concentration in addition to a person's individual susceptibility. Small amounts may be of no health consequence, whereas larger amounts may cause adverse health effects." See Centers for Disease Control and Prevention, "Executive Summary," *National Report on Human Exposure to Environmental Chemicals*, 2005; www.cdc.gov/ exposurereport/.

37   **D. Richard Jackson:** Nena Baker, *The Body Toxic: How the Hazardous Chemistry of Everyday Things Threatens Our Health and Well-Being* (New York: North Point Press, 2008), p. 25.

When it comes to scientific debates: Not surprisingly, industry's push-back against body burden studies has been intense. "The Environmental Working Group: Peddlers of Fear," shouted the head-line on a release put out in January 2004 by the Capital Research Center, an industry group that "analyzes organizations that promote the growth of government." The group accused the EWG of specializ-ing in "junk science" to "foment health scares about various foods, pes-ticides and other products." The essay accuses the group of being a "Trojan Horse for the Organic Food industry," and says its attacks on synthetic chemicals are similar to those made in Rachel Carson's *Silent Spring*, which it calls "a misconceived attack on man-made chemicals."

"Pesticides are a testimony to human ingenuity in the service of the most elementary human need: providing food to eat," it contin-ues. "There is no doubt that pesticides are one of the many innova-tions that have dramatically increased human life expectancy over the past few decades. But pesticides are EWG's enemy."

The article then makes a clever turn. By promoting pesticide-free agriculture, it argues, EWG is pushing the cost of cancer-fighting fruits and vegetables beyond the reach of "people of moderate or lower incomes." Advocating for organic food, in other words, is "a sure way to raise the cancer risk of millions of people."

"Because EWG uses its junk science–laden reports as fodder for class-action lawsuits, the organization has achieved an influence that far exceeds many better-known environmental organizations," the report maintains. "But by focusing on a narrow range of hot topics and by playing fast and loose with the facts, EWG has proven adept at 'turning raw data into useable information.' It's an effective voice for an irresponsible cause." The report, by Bonner R. Cohen, can be found at www.capitalresearch.org.

38   **Lauralee was not the only:** "Body of Evidence: A Study of Pollution in Maine People," Alliance for a Clean and Healthy Maine, June 2007; available at www.cleanandhealthyme.org.

42   **The chemical industry chimed in:** Josie Huang, "Study Finds 36 Toxic Chemicals in Bodies of Mainers Tested," *Portland Press Herald*, June 12, 2007.

46 **As word got out:** Doug Harlow, "Starks: Demise of Birds Probed," Kennebec *Morning Sentinel*, December 18, 2008.

47 **Since it was invented:** Rebecca Renner, "It's in the Microwave Popcorn, Not the Teflon Pan," *Environmental Science and Technology News*, November 16, 2005. In recent years, worries have cropped up over our remarkable exposure to the compounds, and the links between the chemicals and cancer and birth defects. Virtually every American is thought to have PFOAs in their blood, and by 2000, 3M was producing eight million pounds of PFOS (another perfluorinated chemical) a year. Unfortunately, the stuff was not staying put in carpets and pans. Researchers have begun finding PFOAs and PFOSs in the blood of everything from polar bears in the Arctic to cormorants in the Sea of Japan. DuPont recently paid $50 million in cash and $22 million in legal fees to settle a lawsuit brought by workers at a plant in West Virginia who claimed Teflon chemicals contributed to birth defects and other illnesses. (This raises an interesting etymological question: What would it mean, in this day and age, to refer to Ronald Reagan as the Teflon president?)

Under pressure from the EPA, 3M agreed to phase out PFOSs in 2000. Six years later, DuPont, 3M, and six other chemical companies signed a nonbinding, voluntary agreement with the EPA to reduce PFOAs by 95 percent by 2010, but the compounds are still being used to make Teflon pans and Gore-Tex clothing—the last of which gives pause to an outdoor enthusiast like Amy Graham. See John Butenoff, "The Applicability of Biomonitoring Data for Perfluorooctanesulfonate to the Environmental Public Health Continuum," *Environmental Health Perspectives*, vol. 114 (2006), pp. 1776–82; B. H. Alexander et al., "Mortality of Employees of a Perfluorooctanesulphonyl Fluoride Manufacuring Facility," *Occupational Environmental Medicine*, vol. 60 (2003), pp. 722–29; and Chris Summers, "Teflon's Sticky Situation," BBC News, October 7, 2004.

48 **But these little dolls:** One phthalate, DEHP—di(2-ethylhexyl) phthalate—was used in baby pacifiers and plastic food wrap until it was shown to be a probable carcinogen. It is still added to PVC pipe, to give it flexibility. But again, as with flame retardants, residues of DEHP have been found in food with high fat content, like eggs, milk, cheese, margarine, and seafood. See Sandra Steingraber, *Living*

Downstream: An Ecologist Looks at Cancer and the Environment (Reading, Mass.: Addison-Wesley, 1997), p. 112.

CHAPTER TWO: THE HOME

54   **How often do you use:** "Study Finds Pesticide Link to Childhood Leukemia," Agence France-Press, July 29, 2009.
      Cancer, it is true: Theo Colburn, Dianne Dumanoski, and John Peterson Myers, *Our Stolen Future* (New York: Plume Books, 1996), p. 40.

56   **Donnay liked all of this:** Judith Schreiber et al., "Apartment Residents' and Day Care Workers' Exposures to Tetrachloroethylene and Deficits in Visual Contrast Sensitivity," *Environmental Health Perspectives*, vol. 110, no. 7 (July 2002), pp. 655–64.

57   **Then we got to the bathrooms:** D. Andrew Crain et al., "An Ecological Assessment of Bisphenol-A: Evidence from Comparative Biology," *Reproductive Toxicology*, vol. 24 (2007), pp. 225–39; see also Sheela Sathyanarayana et al., "Baby Care Products: Possible Sources of Infant Phthalate Exposure," *Pediatrics*, vol. 121, no. 2 (February 2008).
      In recent years: Some of these problems have begun showing up in the adult male populations as well. A 2000 analysis of more than one hundred clinical studies done between 1934 and 1996 by Dr. Shanna Swan, then at the University of Missouri, found that average sperm counts in Europe were dropping 3 percent a year; in the United States, it was 1.5 percent a year. The study found no such drop in nonindustrialized countries. The incidence of testicular cancer has also increased "significantly" in Caucasians in Western countries. Yet while the damage may be done to the male baby, the toxic chemicals are delivered through the bodies of the contaminated mothers. The CDC's 2000 study found one phthalate, diethyl phthalate (DEP), in the bodies of every one of the women of child-bearing age. See Swan et al., "The Question of Declining Sperm Density Revisited: An Analysis of 101 Studies Published 1934–1996," *Environmental Health Perspectives*, vol. 108, no. 10 (October 2000).

58   **Public health experts:** Nathan Welton, "Embalming Toxins," *E: The Environmental Magazine*, March–April 2003. See also John A.

McLachlan et al., "Endocrine Disruptors and Female Reproductive Health," *Best Practice & Research Clinical Endocrinology & Metabolism*, vol. 20, no. 1 (March 2006), pp. 63–75, and Jane Houlihan, Charlotte Brody, and Bryony Schwan, "Not Too Pretty," Environmental Working Group, July 8, 2002.

69 **It is true that MCS:** Pamela Reed Gibson, *Multiple Chemical Sensitivity: A Survival Guide* (Oakland, Calif.: New Harbinger Publications, 2000), pp. 9–10, 35.

70 **"We are a culture":** Gibson, *Multiple Chemical Sensitivity*, p. 163.

71 **In the mid-1990s:** For more information about Donnay's organization, see www.mcsrr.org.

   But it's not just air pollution: "FEMA Runs for Cover," unsigned editorial, *New York Times*, July 22, 2007.

72 **And so it is:** K. Jakobsson, K. Thuresson, L. Rylander, A. Sjodin, L. Hagmar, and A. Bergman, "Exposure to Polybrominated Diphenyl Ethers and Tetrabrombisphenol A Among Computer Technicians," *Chemosphere*, vol. 46, no. 5 (2002), pp. 709–16. See also A. Sjodin, L. Hagmar, E. Klasson-Wehler, K. Kronholm-Diab, E. Jakobsson, and A. Bergman, "Flame Retardant Exposure: Polybrominated Diphenyl Ethers in Blood from Swedish Workers," *Environmental Health Perspectives*, vol. 107, no. 8 (August 1999), pp. 643–48; Hale et al., "Potential Role of Fire-Retardant-Treated Polyurethane Foam as a Source of Brominated Diphenyl Ethers to the US Environment," *Chemosphere*, vol. 46, no. 5 (2002), pp. 729–35; Bryony Wilford et al., "Passive Sampling Survey of Polybrominated Diphenyl Ether Flame Retardants in Indoor and Outdoor Air in Ottawa, Canada: Implications for Sources and Exposures," *Environmental Science and Technology*, vol. 38 (2004), pp. 5312–18; and Ruthann Rudel et al., "Phthalates, Alkylphenols, Pesticides, Polybrominated Diphenyl Ethers, and Other Endocrine-Disrupting Compounds in Indoor Air and Dust," *Environmental Science and Technology*, vol. 37, no. 20 (2003), pp. 4543–53.

73 **The EPA has warned:** Stephanie Desmon, "The Danger Inside," *Baltimore Sun*, March 2, 2009, and Mary H. Ward et al., "Proximity to Crops and Residential Exposure to Agricultural Herbicides in Iowa," *Environmental Health Perspectives*, vol. 114, no. 6 (June 2006).

74 **In 2003, a team of researchers:** Ruthann Rudel et al., "Phthalates, Alkylphenols, Pesticides, Polybrominated Diphenyl Ethers, and

Other Endocrine-Disrupting Compounds in Indoor Air and Dust," *Environmental Science and Technology*, vol. 37, no. 20 (2003), pp. 4543–53.

74   **And the human inhabitants:** National Public Radio, *All Things Considered*, April 17, 2008.

CHAPTER THREE: THE BIG BOX STORE

82   **Which, I suppose:** "Top 10 U.S. Retailers: Walmart, Kroger and Costco Lead List," *Retail Info Systems News*, December 15, 2009; available online at www.risnews.com.

88   **And in people?:** In 2002, New York's attorney general released a report showing that commercial pesticide workers had applied 4.1 million pounds of dry pesticides and 820,000 gallons of liquid pesticides to apartments, schools, parks, day care centers, senior centers, hospitals, offices, and office buildings across New York State. Home owners, landlords, and apartment dwellers purchased many additional thousands of pounds of pesticides for private application. The pesticides used in New York include carcinogens, endocrine disruptors, chemicals capable of causing birth defects, and chemicals that can cause brain damage. Among the findings: none of the schools surveyed warned parents or students about pesticide applications; and illegal or restricted pesticides were frequently for sale and in wide use, even by schools.

The attorney general's survey discovered that some very dangerous pesticides are used by governmental agencies in places where people live, work, and play. For example, the New York City schools, the Syracuse schools, and the Albany Housing Authority used hydramethylnon, a pesticide classified by the U.S. EPA as a "possible carcinogen." The Syracuse Housing Authority used Baygon, a "probable carcinogen." All of these chemicals are applied in areas frequented by infants and children. By reducing or eliminating toxic pesticides, New York could save billions of dollars in health care costs each year, the report said.

See Michael H. Surgan, Thomas Congdon, et al., "Pest Control in Public Housing, Schools and Parks: Urban Children at Risk," Environmental Protection Bureau, Office of the New York State Attorney General, August 2002. The full report can be found at

www.oag.state.ny.us/bureaus/environmental/pdfs/pest_control_pu
blic_housing.pdf.

89   **Next to this was a can:** See www.epa.gov/iaq/voc.html.

92   **Which, of course, is just:** Manuela Gago-Dominguez et al., "Use of
Permanent Hair Dyes and Bladder Cancer Risk," *International
Journal of Cancer*, vol. 91 (2001), pp. 575–79. Europe has banned
the use of paradioxane in all personal care products and recently recalled all contaminated products. Not so the United States, where
synthetic ingredients, in the words of one industry spokesman, are
like "salt."

"A little salt on your peas or tomatoes can be good," says John
Bailey, executive vice president for science of the Cosmetic, Toiletry
and Fragrance Association. "But a lot of salt can have adverse health
effects on your blood pressure, and too much can be fatal." See
Devra Davis, "A Thousand Threats: Cancer-Causing Chemicals
Don't Work Alone, but in Tandem," *Newsweek International*, March
5, 2007, and Natasha Singer, "Skin Deep: Should You Trust Your
Makeup?," *New York Times*, February 15, 2007.

For decades, the health effects: See news.google.com/
newspapers?nid=1356&dat=19620211&id=UpEUAAAAIBAJ&sjid
=MwUEAAAAIBAJ&pg=6077,1540065. See also Nancy Irwin
Maxwell, "Social Differences in Women's Use of Personal Care
Products: A Study of Magazine Advertisements, 1950–1994," Silent
Spring Institute, November 1, 2000.

93   **But in recent years:** Zhu Hao and John Xiao, "Uptake, Translocation
and Accumulation of Manufactured Iron Oxide Nanoparticles by
Pumpkin Plants," *Journal of Environmental Monitoring*, vol. 10
(2008), pp. 713–17.

A couple of years ago: Stacey Malkan, *Not Just a Pretty Face: The
Ugly Side of the Beauty Industry* (Gabriola Island, B.C.: New Society
Publishers, 2007), pp. 48, 95–96. The campaign targeted companies
like Los Angeles–based OPI Products, the largest maker of nail polish and nail treatment products in the world. OPI is known for off-
beat products like Aphrodite's Nightie, Melon of Troy, and All
Lacquered Up, which are sold in nail salons in seventy countries.
OPI had already taken the phthalate DBP out of its E.U. products,
but refused to remove it, toluene, or formaldehyde—all three of
which are on California's Proposition 65 list of chemicals known to

cause cancer and reproductive toxicity—from cosmetics sold in the United States. In 2004, the group sent a letter signed by more than fifty environmental, health, and women's groups to 250 leading cosmetic companies, asking them to remove phthalates and sign the Compact for Safe Cosmetics, which would lead to the replacement of all toxic ingredients with safer alternatives. There was some good news: within a year, 100 companies had signed the pledge; a year after that, more than 300 companies had signed. By last year, the list had grown to 1,000. Still, most of the major companies have yet to sign, though some have agreed to remove the chemical banned by the E.U.—another indication of market pressure avoided or heeded (and niches found). The campaign took out an ad in 2002 in the *Washington Post* that showed a sultry, fully made-up model and the line "Something has come between me and my Calvins: toxic chemicals in beauty products." The ad targeted product lines manufactured by Unilever (including Calvin Klein's Eternity, Aqua Net Hair Spray, and Dove Solid Anti-Perspirant) that contained phthalates. See also safecosmetics.org.

95   **And these troubles can arise:** Sandra K. Ceario and Lisa A. Hughes, "Precocious Puberty: A Comprehensive Review of Literature," *Journal of Obstetric, Gynecologic and Neonatal Nursing*, vol. 36, no. 3 (2007), p. 263. For the full NIH report, see www.niehs .nih.gov/health/topics/agents/endocrine/docs/endocrine.pdf. See also Diana Zuckerman, "When Little Girls Become Women: Early Onset of Puberty in Girls," *Ribbon*, a newsletter of the Cornell University Program on Breast Cancer and Environmental Risk Factors in New York State, vol. 6, no. 1 (2001); Maryann Donovan et al., "Personal Care Products That Contain Estrogens or Xenostrogens May Increase Breast Cancer Risk," *Medical Hypotheses*, vol. 68 (2007), pp. 756–66; and "Ten-Year-Old Bravely Battles Breast Cancer," CBS News, May 19, 2009.

All of this: Theo Colburn, Dianne Dumanoski, and John Peterson Myers, *Our Stolen Future* (New York: Plume Books, 1996), p. 240.

97   **There they were again:** "Protect Your Family from the Hidden Hazards in Air Fresheners," Natural Resources Defense Council, September 2007.

Nearly 5 million metric tons: Douglas Fischer, "What's in You," the first of a three-part series in the *Oakland Tribune*, March 10, 2005.

98   **The researchers expressed their frustration:** Sheela Sathyanarayana, Catherine J. Karr, et al., "Baby Care Products: Possible Sources of Infant Phthalate Exposure," *Pediatrics*, vol. 121, no. 2 (February 2008), pp. 260–68.

In state after state: The Washington study can be found at www.pollutioninpeople.org/; the Oregon study can be found at www.oeconline.org/our-work/kidshealth/pollutioninpeople.

99   **In September 2000:** B. C. Blount, M. J. Silva, S. P. Caudill, L. L. Needham, J. L. Pirkle, E. J. Sampson, G. W. Lucier, R. J. Jackson, and J. W. Brock, "Levels of Seven Urinary Phthalate Metabolites in a Human Reference Population," *Environmental Health Perspectives*, vol. 108, no. 10 (October 2000), pp. 979–82. See also M. C. Kohn, F. Parham, S. A. Masten, C. J. Portier, M. D. Shelby, J. W. Brock, and L. L. Needham, "Human Exposure Estimates for Phthalates," *Environmental Health Perspectives*, vol. 108, no. 10 (October 2000).

A study of 85 people: Males' perineums at birth are usually about twice as long as those of females, in both humans and laboratory rodents. In this study, the baby boys of women with the highest phthalate exposures were ten times as likely to have a shortened AGD, adjusted for baby weight, as the sons of women who had the lowest phthalate exposures. The length difference was about 20 percent, according to the study, led by Shanna Swan, now at the University of Rochester School of Medicine and Dentistry; the results were published in *Environmental Health Perspectives*. Among boys with shorter AGD, 21 percent also had incomplete testicular descent and small scrotums, compared with 8 percent of the other boys. Some endocrinologists called this the first study to link an industrial chemical measured in pregnant women to altered reproductive systems in offspring. "It is really noteworthy that shortened AGD was seen," said Niels Skakkebaek, a reproductive-disorder expert at the University of Copenhagen. "If it is proven the environment changed the [physical characteristics] of these babies in such an anti-androgenic manner, it is very serious." See Peter Waldman, "From Ingredient in Cosmetics, Toys, a Safety Concern," *Wall Street Journal*, October 4, 2005. See also Shanna Swan et al., "Decrease in Anogenital Distance Among Male Infants with Prenatal Phthalate Exposure," *Environmental Health Perspectives*, vol. 113, no. 8 (August 2005).

99 **Strangely enough, federal guidelines:** Nena Baker, *The Body Toxic: How the Hazardous Chemistry of Everyday Things Threatens Our Health and Well-Being* (New York: North Point Press, 2008), p. 92. Because the Food, Drug, and Cosmetic Act does not require evidence of the safety of ingredients in cosmetics prior to their marketing, the FDA is not compelled to review or approve these products before they go on the market. Instead, the FDA allows manufacturers to (voluntarily) report customer complaints, and only then will the FDA take legal action. And product recalls? Recalls, the FDA reports, "are voluntary actions taken by manufacturers or distributors to remove from the marketplace products that represent a hazard or gross deception, or that are somehow defective."

So once again, instead of effective government oversight, we have voluntary industry self-regulation, in this case something called the Cosmetic Ingredient Review (CIR), an industry panel made up of seven physicians and scientists working for a group financed by the Personal Care Products Council. In its entire history, the CIR has managed to review just 11 percent of some 10,500 ingredients used in cosmetics. Another way of saying this is that 89 percent of the ingredients have never been evaluated.

For the consumer, sifting through industry-sponsored information and independent information can be difficult, even disorienting. And as with other products, this confusion can, at times, seem intentional. If you turn to the FDA, you get this: "Many nail products contain potentially harmful ingredients, but are allowed on the market because they are safe when used as directed. For example, some nail ingredients are harmful only when ingested, which is not their intended use." See also www.cfsan.fda.gov/~dms/cos-206.html; www.cosmeticsdatabase.com/research/; www.cir-safety.org/; and www.cfsan.fda.gov/~dms/cosx-nail.html.

100 **So, once again, we are left:** www.cosmeticsinfo.org/aboutus.php. In 1988, U.S. representative Ron Wyden (D-Oregon) held hearings on Capitol Hill for the fiftieth anniversary of the Food, Drug, and Cosmetic Act. Wyden wanted Congress to require cosmetic firms to test ingredients before they got to market and to give the FDA access to consumer complaints about products. He got nowhere. In 2007, after the Campaign for Safe Cosmetics reported finding lead contamination in 61 percent of the name-brand red lipsticks it

tested, Senators John Kerry, Barbara Boxer, and Dianne Feinstein asked the FDA to take "immediate action" to reduce lead exposure in lipstick. A few weeks later, the agency announced that it had allocated resources for independent testing of some lipsticks, but said it was not valid to compare lead levels in lipstick (for which there is no safe standard) with safety recommendations for lead in candy. The FDA has no authority to "subject cosmetics to pre-market approval," the FDA reported. Lipstick is less of a worry because it is used only topically and "is ingested in much smaller quantities than candy." For the full FDA report, see www.fda.cfsan.fda .gov/~dms/cos-pb.html.

100 **Such rhetoric does not always seem reassuring:** While few studies have been done on nail salons, one report found that nail technicians had greater problems with attention and cognitive processing and that the longer people stayed in the industry, the greater the severity of problems. Another study found an increased risk of spontaneous abortions in cosmetologists in North Carolina; a third found cosmetologists at higher risk for Hodgkin's disease. See Alexandra Gorman and Philip O'Connor, *Glossed Over: Health Hazards Associated with Toxic Exposure in Nail Salons,* a report for Women's Voices for the Earth, February 2007. See also G. L. LoSasso, "Neurocognitive Sequelae of Exposure to Organic Solvents and (Meth)acrylates Among Nail Studio Technicians," *Neuropsychiatry, Neuropsychology and Behavioral Neurology,* vol. 15, no. 1 (March 2002), pp. 44–55; E. M. John et al., "Spontaneous Abortions Among Cosmetologists," *Epidemiology,* vol. 5, no. 2 (1994), pp. 147–55; C. Robinson et al., "Cancer Mortality Among Women Employed in Fast-Growing U.S. Occupations," *American Journal of Industrial Medicine,* vol. 36 (1999), pp. 186–92. In California, an occupational hazard hotline found that manicurists and cosmetologists made the third-greatest number of calls about pregnancy. Many Asian-American nail workers quit their jobs when they get pregnant, to avoid health impacts; for reasons involving their culture and immigration, most are unwilling to come forward about their children's disabilities. See the National Asian Pacific American Women's Forum, www.napawf.org/file/issues/issues-Nail Salon.pdf.

In Boston, researchers found considerable consensus among Vietnamese nail technicians that their job was affecting their health.

Most reported that chemical odors and workplace dust made them feel bad; they specified headaches, respiratory problems, and skin irritation. Many said their symptoms improved when they were away for a couple of days. Few said their employers had given them any health information; if anything, they were advised to wear masks and keep the windows open. At the Nail Stop salon in Newark, Delaware, manager Le Banh said he gives his employees leave if they become pregnant. "If we find out the lady is pregnant, we ask them to leave and then come back because the chemicals are bad for the babies," Banh said. See www.aiha.org/aihce06/handouts/po125roelofs.pdf. See also Maddie Thomas, "Finding a Safer Manicure," University of Delaware *Review*, November 25, 2008.

In 2007, after receiving complaints from Asian-American nail salon workers in and around Houston, the EPA published a guide called *Protecting the Health of Nail Salon Workers* that would seem to be of interest to anyone using nail products. Many of the products used in nail salons—and presumably in private homes—"should be used and handled properly to minimize potential for overexposure." What might some of the problems of overexposure be? Here are a few, according to the EPA: DBP, a phthalate used in nail polish and nail hardener, "may be hazardous to human reproduction and development." Acetone, a nail polish remover and fingernail glue remover, may cause central nervous system depression. Formalin (a compound containing 37 to 50 percent formaldehyde and 6 to 15 percent alcohol stabilizer, used as a nail hardener) "may be a carcinogen if inhaled in high concentrations or for long periods." Toluene, also used to remove nail polish and fingernail glue, "may cause irritation to eyes and nose, weakness, exhaustion, confusion, inappropriate feelings of happiness, dizziness, headache, dilated pupils, runny eyes, anxiety, muscle fatigue, inability to sleep, feeling of numbness/tingling, skin rash, and in more serious cases of overexposure or intentional abuse, liver and kidney damage." Titanium dioxide, used in nail polish and in powder for artificial nails, "may be an occupational carcinogen." The EPA also notes that "some nail products contain liquid methyl methacrylate (MMA)," which is "a poisonous and harmful substance" that has been banned or restricted in "at least 30 states." This, I suppose, is useful to know. But it does raise a question: Why is it still available in the other 20 states? See Enivronmental

Protection Agency, *Protecting the Health of Nail Salon Workers,*
March 2007 (www.epa.gov/dfe/pubs/projects/salon/index.htm).

102 **You'll find similar tags:** See www.cpsc.gov/businfo/frnotices/fr96/
frsleeplg.html.

Did the flame retardants save our lives?: See National Academies
Press, www.nap.edu/openbook.php?record_id=9841&page=1, p. 15.

103 **People have been inventing:** Overall, global demand for flame retardants jumped from 20,000 tons in 1984 to about 70,000 tons in
2004. About 95 percent of all deca PBDEs produced worldwide are
used in the United States and Canada. In 2001, American companies
alone used 66 millions pounds of them. See Anita Lahey, "Unsafe
Assumption," *Canadian Geographic,* May–June 2003. See also Yana
Kucker and Meghan Purvis, "Body of Evidence: New Science in the
Debate Over Toxic Flame Retardants and Our Health," National
Association of State PIRGs and the Environment California
Research and Policy Center, 2004.

The chemical industry maintains that deca is a larger chemical
compound than either penta or octa, and is thus less easily absorbed
into the body than its chemical cousins. But a number of studies dispute this, and have shown that once deca is "exposed"—to sunlight
or the digestive systems of fish, for example—it can break down into
its more dangerous cousins.

These additives are not incidental: T. Madsen, S. Lee, and T. Olle,
"Growing Threats: Toxic Flame Retardants and Children's Health,"
Environment California Research and Policy Center, 2003. See also
R. C. Hale, M. J. La Guardia, E. Harvey, and T. M. Mainor, "Potential
Role of Fire Retardant–Treated Polyurethane Foam as a Source of
Brominated Diphenyl Ethers to the US Environment," *Chemosphere,*
vol. 46, no. 5 (2002), pp. 729–35.

Here's the problem: T. McDonald, "A Perspective on the Potential
Health Risks of PBDEs," *Chemosphere,* vol. 46, no. 5 (2002), pp.
745–55; T. O. Darnerud et al., "Polybrominated Diphenyl Ethers:
Occurrence, Dietary Exposure, and Toxicology," *Environmental
Health Perspectives,* vol. 109, supp. 1 (March 2001), pp. 49–68. See
also P. O. Darnerud, "Toxic Effects of Brominated Flame Retardants
in Man and Wildlife," *Environment International,* vol. 29, no. 6
(2003), pp. 841–53; I. Branchi, F. Capone, E. Alleva, and L. G. Costa,
"Polybrominated Diphenyl Ethers: Neurobehavioral Effects

Following Developmental Exposure," *Neurotoxicology*, vol. 24, no. 3 (2003), pp. 449–62; and T. A. McDonald, "A Perspective on the Potential Health Risks of PBDEs," *Chemosphere*, vol. 46, no. 5 (2002), pp. 745–55; H. Stapleton, M. Alaee, and J. Baker, "Debromination of Decabromodiphenyl Ether by Juvenile Carp (*Cyprinus carpio*)," *Organohalogen Compounds*, vol. 61 (2003), pp. 21–24; M. LaGuardia, R. C. Hale, and E. Harvey, "Are Waste Water Treatment Plants Sources for Polybrominated Diphenyl Ethers?," a paper presented at the Annual Meeting of the Society of Environmental Toxicology and Chemistry (SETAC), 2003; and Watamabe et al., "Formation of Brominated Dibenzofurans from the Photolysis of Flame Retardant Decabromidiphenyl Etherin Hexane Solution by UV and Sunlight," *Bulletin of Environmental Containment Toxicology*, vol. 35 (1987), pp. 953–59.

104 **Flame retardants are one:** National Geographic Society, *The Green Guide: The Compete Reference for Consuming Wisely* (Washington, D.C.: National Geographic, 2008).

105 **Toxic flame retardants were first identified:** Anita Lahey, "Unsafe Assumption," *Canadian Geographic*, May–June 2003. See also Yana Kucker and Meghan Purvis, "Body of Evidence: New Science in the Debate Over Toxic Flame Retardants and Our Health," National Association of State PIRGs and the Environment California Research and Policy Center, 2004.

The quantity of flame retardants: Rebecca Renner, "What Fate for Brominated Flame Retardants?," *Environmental Science and Technology*, vol. 34, no. 9 (May 2000), pp. 223A–26A; A. Bocio, J. M. Llobet, J. L. Domingo, J. Corbella, A. Teixido, and C. Casas, "Polybrominated Diphenyl Ethers (PBDEs) in Foodstuffs: Human Exposure Through the Diet," *Journal of Agricultural and Food Chemistry*, vol. 51, no. 10 (2003), pp. 3191–95; R. C. Hale, M. Alaee, J. B. Manchester-Neesvig, H. M. Stapleton, and M. G. Ikonomou, "Polybrominated Diphenyl Ether Flame Retardants in the North American Environment," *Environment International*, vol. 29, no. 6 (2003), pp. 771–79; D. Meironyte et al., "Analysis of Polybrominated Diphenyl Ethers in Swedish Human Milk: A Time-Related Trend Study, 1972–1997," *Journal of Toxicological Environmental Health, Part A*, vol. 58, no. 6 (1999), pp. 329–41; K. Hooper and J. She, "Lessons from the Polybrominated Diphenyl Ethers (PBDEs):

Precautionary Principle, Primary Prevention, and the Value of Community-Based Body-Burden Monitoring Using Breast Milk," *Environmental Health Perspectives*, vol. 111, no. 1 (2003), pp. 109–13.

105 **Flame retardants are "fat-soluble":** "The exponential increase of flame retardants in breast milk is alarming and calls for measures to stop the exposure," the Swedish report said. "Since also these environmental pollutants are or may become globally spread the consequences are of international concern. The present study shows that monitoring of breast milk serves an important sentinel function in detecting the occurrence and exposure of widespread toxic contaminants at an early stage and provides possibility to take measures before adverse health effects appear." See K. Noren and D. Meironyte, "Certain Organochlorine and Organobromine Contaminants in Swedish Human Milk in Perspective of Past 20–30 Years," *Chemosphere*, vol. 40 (2000), pp. 1111–23.

106 **As public concern grew:** Sweden is the only nation with a comprehensive breast milk monitoring program, so it has been difficult to track these trends elsewhere. However, in regions where bans and restrictions have not been established, available studies are showing that concentrations of flame retardants in breast milk have risen far past Sweden's 1997 peak. See P. O. Darnerud, M. Aune, S. Atuma, W. Becker, R. Bjerselius, S. Cnattingius, and A. Glynn, "Time Trend of Polybrominated Diphenyl Ether (PBDE) Levels in Breast Milk from Uppsala, Sweden, 1996–2001," *Organohalogen Compounds*, vol. 58 (2002), pp. 233–36. At Sweden's urging, the European Commission banned the sale and use of penta and octa in 2004, and began phasing out the use of deca in electronics a couple of years later. The International POPs Elimination Network, an alliance of nongovernmental organizations from sixty-five countries, is working to get all three added to the "dirty dozen" chemicals listed by the Stockholm Convention, a global treaty designed to protect people and the environment from toxic chemicals. For more information about the International POPs Elimination Network, see www.ipen.org/ or the European Union's Directive 2002/95/EC of the European Parliament and of the Council, "on the restriction of the use of certain hazardous substances in electrical and electronic equipment," Document PE-CONS 3662/02, Brussels, November 8, 2002, available at europa.eu.int/.

In this country, no federal regulatory action has been taken to ban or restrict flame retardants. The EPA reached a voluntary agreement with Great Lakes Chemical Corporation, the compounds' only manufacturer, to cease production of penta and octa in 2004, but products containing the stuff still abound—both in older products and in new ones imported from places like China. As with other toxins, laws restricting their manufacture in the United States do very little if most of the products are coming in from other places. Some of the chemicals replacing PBDEs are themselves suspect. Foam once made with penta is now made with chlorinated tris, a chemical banned from children's pajamas in the 1970s because it was shown to cause cancer.

In the summer of 2008, reporters from the *Milwaukee Journal Sentinel* examined the EPA's oversight of tris, which has also been linked to reproductive and developmental problems. The paper found that the EPA's program set up to warn the public about toxic chemicals relies on studies that were at best decades old, and at worst had been paid for by the very industries they were supposed to oversee. "This lopsided assessment is the latest example of how the EPA gives preferential treatment to the chemical industry," the newspaper reported. "Instead of conducting independent reviews of chemicals, the EPA allows chemical manufacturers to characterize the safety of the products they make. The EPA then posts those claims on its Web site, often without verifying or correcting the information."

Arlene Blum, a scientist from the University of California–Berkeley whose work led to the chemicals being taken out of children's sleepwear in the 1970s, told the paper she was astonished to learn that chlorinated tris is back in such widespread use in other consumer products, particularly couches and places where children play. "We are going from one toxin to another with no requirement to tell people about the threats to their health and safety," Blum said. "It's been more than 30 years and the chemical industry hasn't bothered to come up with an alternative? We can't do any better than this?" See Susanne Rust and Meg Kissinger, "Hazardous Flame Retardant Found in Household Objects," *Milwaukee Journal Sentinel*, July 13, 2008.

106 **So how flame retardant:** Arnold Schecter et al., "Polybrominated

Diphenyl Ethers (PBDEs) in U.S. Mothers' Milk," *Environmental Health Perspectives*, vol. 111, no. 14 (November 2004).

106 **And if you think those women:** Kellyn S. Betts, "A New Record for PBDE's in People," *Environmental Science and Technology*, May 25, 2005; pubs.acs.org/subscribe/journals/esthag-w/2005/may/science/kb_newrecord.html. The full study can be found at pubs.acs.org/cgi-bin/article.cgi/esthag/2005/39/i14/html/es050399x.html.

107 **As you can see:** Half the salmon sold worldwide, most of it Atlantic salmon, is raised on farms in northern Europe, Chile, Canada, and the United States. The team tested seven hundred fish taken from the top-producing salmon farms in the world: in Norway, Chile, Scotland, British Columbia, eastern Canada, the Faroe Islands, and Washington State. The highest levels were found in fish farmed in Europe; the lowest levels were in farmed salmon from Chile. Most of the salmon sold in European stores come from European farms, which produce the most contaminated fish; most of the salmon in the United States comes from Chile and Canada. Among the wild fish, PBDE concentrations were highest in wild chinook from British Columbia, perhaps because, as fish eaters themselves, they feed higher on the food chain than other salmon, which tend to eat smaller invertebrates and zooplankton. "It has been suggested that PBDE concentrations now observed in humans may leave little or no margin of safety," the study's authors write. "Thus, prudent public health practice argues for the selective consumption of food, including many wild salmon species that contain comparatively lower concentrations of PBDEs." See Ronald Hites et al., "Global Assessment of Organic Contaminants in Farmed Salmon," *Science*, vol. 303 (January 9, 2004), pp. 226–92; see also Ronald Hites, Jeffrey Forman, et al., "Global Assessment of Polybrominated Diphenyl Ethers in Farmed and Wild Salmon," *Environmental Science and Technology*, vol. 38 (2004), pp. 4945–49. The salmon farmers' association has a lot to lose if consumers begin to believe that farmed salmon is not safe food. British Columbia, Canada's largest aquaculture-producing province, generated sales of C$329.6 million in 2002, up 12.3 percent from 2001; "Chemical Flame Retardants Found in Farmed Salmon," *Environmental News Service*, August 10, 2004. *Managed Care Weekly*, which provides "essential management information to healthcare executives," circulated an article titled

"Salmon Study Shows Levels of PBDEs Pose Little Risk." See *Managed Care Weekly*, September 6, 2004.

107 **All of this news:** Arnold Schecter et al., "Polybrominated Diphenyl Ethers Contamination of United States Food," *Environmental Science and Technology*, vol. 38 (2004), pp. 5306–11. See also "PBDEs and the Environmental Intervention Time Lag," *Environmental Science and Technology Science News*, October 15, 2004. For studies in flame retardants and fish consumption in Japan and Europe, see S. Ohta et al., "Comparison of Polybrominated Diphenyl Ethers in Fish, Vegetables, and Meats and Levels in Human Milk of Nursing Women in Japan," *Chemosphere*, vol. 46, no. 5 (2002), pp. 689–96; Bocio et al., "Polybrominated Diphenyl Ethers (PBDEs) in Foodstuffs: Human Exposure Through the Diet," *Journal of Agricultural and Food Chemistry*, vol. 51, no. 10 (2003), pp. 3191–95; M. Zennegg, M. Kohler, A. C. Gerecke, and P. Schmid, "Polybrominated Diphenyl Ethers in Whitefish from Swiss Lakes and Farmed Rainbow Trout," *Chemosphere*, vol. 51, no. 7 (2003), pp. 545–53.

109 **Just two years before:** James E. McWilliams, "Our Home-Grown Melamine Problem," *New York Times*, November 17, 2008.

In May, almost a million tubes: Walt Bogdanich, "From China to Panama, a Trail of Poisoned Medicine," *New York Times*, May 6, 2007. See also Associated Press, "2 Companies Charged over Tainted Toothpaste," *New York Times*, March 6, 2008, and Louise Story and Geraldine Fabrikant, "4 Executives Are Charged over Tainted Toothpaste," *New York Times*, March 7, 2008.

110 **A burden indeed:** Jad Mouawad, "550,000 More Chinese Toys Recalled for Lead," *New York Times*, September 27, 2007.

111 **The hazards of lead:** Ellen Silbergeld, Michael Waalkes, and Jerry M. Rice, "Lead as a Carcinogen: Experimental Evidence and Mechanisms of Action," *American Journal of Industrial Medicine*, vol. 38 (2000), pp. 316–23.

Despite scientific warnings: Melissa Hendricks, "Lead's Nemesis," *Johns Hopkins Magazine*, April 2000. See also Liz Szabo, "Where Does Lead Go? Into Bones," *USA Today*, October 28, 2007. And it's not just kids or the elderly who are at risk when they put lead-tainted things in their mouth. In Germany, lead poisoning even became a concern for pot smokers. Apparently, dealers were adding lead pellets—some of them big enough to see—to increase the weight of bags of their wares. This turned out nicely for the dealers, whose bags weighed 10 percent more and cleared an extra $682 per pound. But when smokers lighted up, the heat of their joints, which can reach 2,200 degrees Fahrenheit, was plenty hot enough to melt the lead and cause respiratory problems. When 145 smokers showed up for a screening at a German clinic, 95 of them had lead poisoning. To the smokers who complained that they thought they were buying an all-natural product, their doctor in Leipzig had one thing to say: "How naïve!" See Denise Grady, "Germany: Marijuana Smokers Were Poisoned with Lead in Leipzig," *New York Times*, April 15, 2008.

112 **Dr. Silbergeld, whose work figured:** Ellen Silbergeld, "The Unbearable Heaviness of Lead," *Bulletin of the History of Medicine*, vol. 77, no. 1 (2003), pp. 164–71. See also Gerald Markowitz and David Rosner, "Cater to the Children: The Role of the Lead Industry in a Public Health Tragedy," *American Journal of Public Health*, vol. 90 (2000), pp. 36–46; Mark Lustberg and Ellen Silbergeld, "Blood Levels and Mortality," *Archives of Internal Medicine*, vol. 162 (November 25, 2002); and Frank Roylance, "Lead Tied to Criminal Behavior," *Baltimore Sun*, May 28, 2008.

More recently, Dr. Silbergeld: Liz Szabo, "Where Does Lead Go? Into Bones," *USA Today*, October 28, 2007; Melissa Hendricks, "Lead's Nemesis," *Johns Hopkins Magazine*, April 2000.

113 **Here, at least in part:** See the company website at www.rc2corp .com/company/multicheck.asp.

In the Frequently Asked Questions: See recalls.rc2.com/recalls _faqs.html.

114 **A few hours later:** See the ASTM website: www.astm.org/ABOUT/ aboutASTM.html.

What federal oversight there is: In 1983, the Consumer Product Safety Commission determined that substantial exposure to DEHP, a common phthalate, could put children at risk for cancer, but the CPSC didn't issue a regulation. Instead, it reached an agreement with the Toy Industry Association to keep DEHP out of pacifiers, rattles, and teethers. The agreement left unregulated all other toys that babies put in their mouths. When President Bush came to office in 2000, he named as head of the consumer protection agency

Harold D. Stratton, a former attorney general of New Mexico. Stratton was a conservative Republican and a Bush campaign volunteer who had made a name for himself by criticizing other law enforcement agencies that tried "to impose their own antibusiness, pro-government regulation views." Later, he became cofounder of a nonprofit group, the Rio Grande Foundation, which claimed to promote "individual freedom, limited government, and economic opportunity." Consumer advocates also criticized Bush for leaving a seat open on the commission's three-seat board, in effect preventing the agency from issuing new rules or penalizing companies that violate existing rules. See Jane Kay, "Toxic: San Francisco Prepares to Ban Certain Chemicals in Products for Tots, but Enforcement Will Be Tough—and Toymakers Question Necessity," *San Francisco Chronicle*, November 19, 2006. See also Stephen Labaton, "Senate Votes to Strengthen Product Safety Laws," *New York Times*, March 7, 2008.

115 **Not only that:** David Barboza, "As More Toys Are Recalled, Trail Ends in China," *New York Times*, June 18, 2007.

Here's what else it looks like: Eric Lipton, "Safety Agency Faces Scrutiny Amid Charges," *New York Times*, September 2, 2007. Joel Ticknor, a professor of environmental health at the University of Massachusetts–Lowell, considers federal protection against toxins in toys to be "weak at best and dysfunctional at worst. Consumers would be astonished if they knew that federal laws regulating chemicals in children's toys all require balancing the benefits of protecting children with the costs to industry of implementing safer alternatives." See Jane Kay, "Toxic: San Francisco Prepares to Ban Certain Chemicals in Products for Tots, but Enforcement Will Be Tough—and Toymakers Question Necessity," *San Francisco Chronicle*. November 19, 2006; see also Jane Kay, "Supervisors Tweak 'Toxic' Child Products; City Will Test up to 100 Items a Year, List Those Illegal to Sell," *San Francisco Chronicle*, April 11, 2007.

116 **In the absence of adequate disclosure:** In the last couple of years alone, Denver's *Rocky Mountain News* has closed up shop, the *Seattle Post-Intelligencer* has stopped publishing a paper, in favor of a Web-only publication, and such major papers as the *Atlanta Journal-Constitution, Baltimore Sun, Philadelphia Inquirer,* and *Boston Globe* have been laying off reporters in droves. But the demise of the press

is a story for another time. Thankfully, for now at least, there are still a few reporters looking into contaminated consumer products.

116 **Around this time:** Contacted by the *Tribune*, the Toysmith Group, an Auburn, Washington–based company that imports the Godzillas from China, said its own tests had found lead levels well under the safety limit. "I don't know what more we could have done to be certain that we are only selling safe products," a lawyer for the company said. Asked about the findings, the spokeswoman for the safety commission, Julie Vallese, said it "always welcomes credible information to be reported to the agency so that we could follow up to determine if in fact toys, based on our own scientific testing, are in violation of the law." See Ted Gregory and Sam Roe, "Hidden Hazards: Kids at Risk; Many More Toys Tainted with Lead, Inquiry Finds," *Chicago Tribune*, November 18, 2007.

117 **The *Tribune* found:** The paper's findings differed from the company's. Baby Einstein is a subsidiary of the Walt Disney Company, and the toy's distributor, Kids II Inc., pays Disney to use the Baby Einstein name. "We realize that the current attention to toy safety may require new discussions and perhaps new industry approaches to testing, quality control and manufacturing," Jeff Cornelison, Kids II's senior vice president of sales and marketing, told the paper. "Kids II will always remain part of those discussions to ensure that our industry continues to provide safe, quality products for children and their families."

Kids II recalled about 35,000 of the blocks sold in the summer of 2007 because of high lead levels. But the company recalled the set because of its painted blue blocks—not the yellow ones tested by the newspaper. The blocks the *Tribune* bought were not covered by the recall. When the paper notified Walgreens of its findings, the store pulled the item nationwide. Company spokeswoman Carol Hively said Walgreens believes the item is not technically a toy and therefore not subject to lead safety rules. But, she said, "a child might think they kind of look like a toy or for the holidays they might be displayed in the reach of a child, so . . . we pulled them. We just wanted to do the right thing." See Ted Gregory and Sam Roe, "Hidden Hazards: Special Report: Kids at Risk," *Chicago Tribune*, November 18, 2007.

All this news: "Consumer Groups Applaud President for Signing

Strong Products Safety Bill into Law," press release from the
Consumer Federation of American, August 14, 2008;
www.consumerfed.org/pdfs/POTUS_Signing_Law_press_release_8
_14_08.pdf. Critics of the bill were many, and fierce. Republican
senator Jim DeMint of South Carolina complained that the bill
would provide a "playground for plaintiffs' attorneys." DeMint also
said that allowing whistle-blower status "makes it legally impossible
to fire disruptive employees" and that the database would be used
"to anonymously smear companies" by circulating "frivolous complaints filed by left-wing interest groups." Giving power to states
"undermines a cooperative relationship between businesses and the
Consumer Product Safety Commission," he added. See Stephen
Labaton, "Senate Votes to Strengthen Product Safety Laws," *New
York Times*, March 7, 2008.

Jim Neill, a spokesman for the National Association of
Manufacturers, claimed that the database would allow "rumor and
innuendo" to smear safe products. The U.S. Chamber of Commerce
was upset that that law would leave room for states to pass safety
standards stricter than the federal guidelines—as Maine, for
example, did with flame retardants. "Manufacturers are going to
have a difficult time because they're going to have a patchwork of
laws to deal with," said Thomas Myers, an attorney for the chamber.
"Theoretically, they can have 50 different laws their products have
to comply with." See Jim Tankersley and Patricia Callahan, "Bill
Targets Toy Safety," *Chicago Tribune*, July 29, 2008.

117 **During the back-and-forth:** Patricia Callahan and Amanda Erickson,
"The Mattel Loophole: Congress May Back Off Pledge of
Independent Toy Testing," *Chicago Tribune*, June 25, 2008. The Bush
White House, always sympathetic to industry concerns, did not want
to extend whistle-blower status to employees who disclosed company information. The administration also opposed the database of
consumer complaints and, taking the Mattel line, requiring testing
laboratories to be independent and privately owned. "These provisions threaten to burden American consumers and industry in un-
productive ways, and may actually harm a well-functioning product
safety system," an administration statement said. See also Ruth
Mantell and Matt Andrejczak, "Manufacturers Resist Safety-Reform
Provisions," *MarketWatch*, March 3, 2008.

118 **The Mattel Loophole:** The battle over the law did not end with
President Bush's signature. Initially, the ban did not cover products
already in company warehouses or already on store shelves, because
manufacturers complained that they would have to dump hundreds
of millions of products to comply. This fact did not sit well with consumer advocates. "How will parents know whether the rubber ducky
they're buying was made today and not in March?" said Rachel
Weintraub, director of product safety for the Consumer Federation of
America. Finally, in February 2009, a federal judge in Manhattan
ruled that the CPSC may not let these phthalate-containing toys remain on shelves. The safety commission will not appeal the ruling,
which "provides unequivocally and unambiguously that no covered
products may be sold as of Feb. 10, 2009," said the ruling. And from
the Department of Unintended Consequences: the American Library
Association is now worried what the new law might mean for school
and public libraries. Since the new law could feasibly apply to ordinary, paper-based children's books—many of which are made with
both phthalates and lead paint—all children's books on public and
school library shelves could, theoretically, be required to be removed
for testing. If not, kids twelve and under would be banned from visiting. See www.cpsc.gov/ABOUT/Cpsia/faq/108faq.html. See also
Annys Shin, "Some Toys with Banned Plastics Will Stay on Market,"
Washington Post, November 19, 2008; "Toys with Banned Toxin Must
Go, Judge Decides," Associated Press, February 5, 2009; and Debra
Lau Whelan, "Consumer Product Safety Improvement Act Delayed
for One Year," *School Library Journal*, February 2, 2009.

121 **What were we to think about the fact:** Frederick S. vom Saal and
Claude Hughes, "An Extensive New Literature Concerning Low-
Dose Effects of Bisphenol A Shows the Need for a New Risk
Assessment," *Environmental Health Perspectives*, vol. 113, no. 8
(August 2005). See also Laura N. Vandenberg et al., "Human
Exposure to Bisphenol A (BPA)," *Reproductive Toxicology*, vol. 24
(2007), pp. 139–77.

122 **Where else can you find BPA?:** Hoa H. Le, Emily M. Carlson, Jason
P. Chua, and Scott M. Belcher, "Bisphenol A Is Released from
Polycarbonate Drinking Bottles and Mimics the Neurotoxic Actions
of Estrogen in Developing Cerebellar Neurons," *Toxicology Letters*,
vol. 176, no. 2 (January 30, 2008), pp. 149–56.

123 **BPA is used to harden:** Jose Antonio Brotons, Maria Fatima Olea-Serrano, Mercedes Villalobos, Vicente Pedraza, and Nicolas Oleanique, "Xenoestrogens Released from Lacquer Coatings in Food Cans," *Environmental Health Perspectives*, vol. 103 (1995), pp. 608–12. The study is available at www.pubmedcentral.nih.gov/articlerender.fcgi?tool=pubmed&pubmedid=7556016. See also Yuji Takao et al., "Release of Bisphenol A from Food Can Lining upon Heating," *Journal of Health Sciences*, vol. 48, no. 4 (2002), pp. 331–34.

"There is a large body": "Hot Liquids Release Potentially Harmful Chemicals in Polycarbonate Plastic Bottles," *University of Cincinnati HealthNews*, January 30, 2008.

Like countless researchers: Hoa H. Le, Emily M. Carlson, Jason P. Chua, and Scott M. Belcher, "Bisphenol A Is Released from Polycarbonate Drinking Bottles and Mimics the Neurotoxic Actions of Estrogen in Developing Cerebellar Neurons," *Toxicology Letters*, vol. 176, no. 2 (January 30, 2008), pp. 149–56. Pharmaceutical researchers initially experimented with BPA in the 1930s because of its hormone-like effects. The compound was passed over in favor of DES (diethylstilbestrol), a more potent estrogenic drug eventually given to pregnant women to prevent miscarriages from the 1940s to the 1970s—resulting in 5 to 10 million exposures. Women exposed to DES are now known to be at increased risk for breast cancer, and their daughters are at increased risk for vaginal and cervical cancer, as well as pregnancy complications and infertility. See Nena Baker, *The Body Toxic*, p. 150.

124 **"As a mother":** Feliza Mirasol, "The BPA Defense," *ICIS Chemical Business*, June 2, 2008. The rhetorical battle over the safety of BPA has been intense. In 2004, the American Plastics Council funded a study by the Harvard Center for Risk Analysis that concluded that the evidence for low-dose effects of BPA were weak. The study was delayed for two and half years—during which time a number of studies showed that the opposite was true. In 2005, Frederick vom Saal, from the University of Missouri, and Claude Hughes, from East Carolina University, surveyed 115 studies done on BPA. "Our current conclusion that widespread exposure to BPA poses a threat to human health directly contradicts several recent reports from individuals or groups associated with or funded by chemical corpora-

tions," they noted. They found that 94 of the studies showed "significant" effects; 31 showed the effects at exposure levels below doses considered "safe."

"Nonetheless, chemical manufacturers continue to discount these published findings because no industry-funded studies have reported significant effects of low doses of BPA," Vom Saal and Hughes said, even though more than 90 percent of government-funded studies have reported significant effects. See Frederick S. vom Saal and Claude Hughes, "An Extensive New Literature Concerning Low-Dose Effects of Bisphenol A Shows the Need for a New Risk Assessment," *Environmental Health Perspectives*, vol. 113, no. 8 (August 2005).

124 **In a follow-up report:** Cary Spivak, Meg Kissinger, and Susanne Rust, "Bisphenol A Is in You," *Milwaukee Journal Sentinel*, December 2, 2007, and Meg Kissinger and Susanne Rust, "Tests Find Chemical After Normal Heating of 'Microwave Safe' Plastics," *Milwaukee Journal Sentinel*, November 15, 2008. See also Marla Cone, "Public Health Agency Linked to Chemical Industry," *Los Angeles Times*, March 4, 2007. That summer, a panel of the world's leading BPA experts, convening in Chapel Hill, North Carolina, concluded that the compound was "a great cause for concern," especially given its potential links to prostate and breast cancer, low sperm counts, and genital abnormalities in male babies. Nonetheless, in November, a panel working for the Bush administration's National Toxicology Program ignored much of the Chapel Hill warnings; its own tepid report declared BPA only of "some concern." If "some concern" sounds like a particularly ambivalent expression, it seems to have been designed that way. The NTP uses a scale ranging from "negligible concern" to "serious concern" with "some concern" lying somewhere in the middle—a hint of a warning, it seems, but not enough to anger industry. Scientists, public health officials, and environmental advocates were angered by the opaque language. See Frederick vom Saal et al., "Chapel Hill Bisphenol A Expert Panel Consensus Statement: Integration of Mechanisms, Effects in Animals and Potential to Impact Human Health at Current Levels of Exposure," *Reproductive Toxicology*, vol. 24, no. 2 (August–September 2007), pp. 131–38.

125 **In May 2008:** Martin Mittelstaedt, "The Hidden Chemical in Cans,"

Toronto *Globe and Mail*, May 29, 2008; Martin Mittelstaedt, "BPA in Cans Safe: Health Canada," Toronto *Globe and Mail*, May 30, 2008. In the spring of 2008, after reports arose showing traces of BPA in the plastic linings of cans of infant formula, Congress asked the FDA how it had decided that there was "no safety concern at the current exposure level." The FDA confessed that it had based its ruling on a pair of industry studies, one of which had never even been published. John Dingell, the chair of the House Committee on Energy and Commerce, wrote a scathing letter to the commissioner of the FDA. "Given that there are dozens of published, peer-reviewed studies related to BPA, your development of critical public health policy in this manner, especially as related to infants and children, seems highly questionable." To read Representative Dingell's letter to Andrew Eschenbach, the commissioner of the FDA, see energycommerce.house.gov/Press_110/110-ltr.040408 .FDA.ltrvonEschenbach.BPA.pdf. In January 2010, the FDA declared it had "some concern about the potential effects of BPA on the brain, behavior and prostate gland of fetuses, infants and children" and would join other federal health agencies in studying the chemical in both animals and humans. See Denise Grady, "F.D.A. Concerned About Substance in Food Packaging," *New York Times*, January 15, 2010. For a timeline of the BPA controversy, see the Environmental Working Group's chronology at www.ewg.org/ reports/bpatimeline.

125 **"In Canada, at least"**: "The general public has a horrible time trying to figure out what is polycarbonate, what is ethylene-based, propylene-based plastics," said Frederick vom Saal, a reproductive scientist at the University of Missouri and one of the country's leading bisphenol A researchers. "They have a terrible time figuring out the chemicals in plastics, because there is no requirement from the FDA to label that. But if it's hard and clear, and it's a baby bottle, or it's a sport water bottle, it's likely polycarbonate, and I use none of those products. This is terrible to impose that kind of burden on the consumer. This is an indirect food additive. Why not have product labeling so that people don't have to suffer like this in terms of stress of, is this product safe or not safe?" See "BPA Exposure and Human Health Concerns," a broadcast of *The Diane Rehm Show*, WAMU, April 29, 2008.

125 **"I eat nothing out of cans":** "BPA Exposure and Human Health Concerns," *The Diane Rehm Show*, WAMU, April 29, 2008.

127 **Even before the jug of blench reaches your home:** David Nitkin, "Deadly Cargo to Roll On," *Baltimore Sun*, May 25, 2008.

129 **And so it goes:** Recent research into the marketing of everything from cigarettes to plastics reveals that companies don't just manufacture products, they manufacture doubt. "Doubt is our product since it is the best means of competing with the 'body of fact' that exists in the minds of the general public," claimed a tobacco industry memo in 1969. "It is also the means of establishing a controversy." See David Michaels, *Doubt Is Their Product: How Industry's Assault on Science Threatens Your Health* (New York: Oxford University Press, 2008), p. 11.

   Our compulsive reliance: Michael Janofsky, "DuPont to Pay $16.5 Million for Unreported Risks," *New York Times*, December 15, 2004; Jack Kaskey, "EPA Says DuPont Withheld Data on Teflon Chemical Risk," *Bloomberg News*, July 11, 2004.

132 **People have been much:** householdproducts.nlm.nih.gov/index .htm; toxnet.nlm.nih.gov/cgi-bin/sis/search/r?dbs+hsdb:@term+ @rn+111-76-2.

133 **When he begins designing:** Take laundry detergent. All detergents are made with "surfactants," compounds that disperse oils and stains. Some synthetic surfactants, like nonylphenol ethoxylates, or NPEs, have become of particular interest to Wolf, especially given recent studies showing their connection to hormone disruption in fish. NPEs are used in pulp and paper processing and by a number of industries, and play a part in the manufacturing of everything from textiles to paints and pesticides. They can also be used as an emulsifier for wax sprayed on fruits and vegetables, as a polymer resin in plastics, and in personal care products like deodorant, skin cream, and shampoo. But the largest volume of NPEs goes into laundry detergents. In 2004 alone, of the more than 260 million pounds of the compound used in the United States, over 80 percent of it went into cleaning products like laundry detergents.

   A survey of the scientific literature on NPEs done by the Sierra Club in 2005 noted that metabolites of NPEs can be found in nearly two-thirds of American streams and that exposure to the compound "causes organisms to develop both male and female sex organs; increases mortality and damage to the liver and kidney; decreases testicular growth and sperm counts in male fish; and disrupts normal male-to-female sex ratios, metabolism, development, growth and reproduction."

In this country, predictably, there is an industry group, this one funded by the soap companies, that is determined to keep the government from restricting its choice of ingredients. The industry has for years been against the general banning of phosphates, the synthetic surfactants once found in virtually all laundry, dishwasher, and liquid soaps. Drained in huge quantities from homes and businesses, phosphates create havoc in rivers and lakes by fertilizing aquatic plants that then suck all the oxygen from the water, creating massive dead zones.

Phosphates were banned in laundry detergents in the 1990s, but Wolf still spends a good bit of his time trying to get state legislators to ban the compounds outright—this in opposition to the political stance of the soap and detergent industry as a whole. Washington State, home to ecologically endangered Puget Sound, banned phosphates in 2010. See the Sierra Club report at www.sierraclub.org/toxics/nonylphenol_ethoxylates3.pdf.

134 **Europe and Latin America:** Leslie Kaufman, "Mr. Whipple Left It Out: Soft Is Rough on Forests," *New York Times*, February 25, 2009.

135 **As companies like Seventh Generation:** The blurring of the marketplace between conventional cleaners and their "alternatives" has left more than a few consumers scratching their heads. Clorox, whose Green Works line is the company's first new brand in twenty years, trumpets an endorsement from an unlikely ally: the Sierra Club. In return for the favor—the first product endorsement in its 116-year history—the environmental group receives an (undisclosed) fee. Such an alliance can be viewed a couple of ways: as a cynical coup for an industry leader trying to gain a foothold in a growing market or as a successful push by an environmental organization to get manufacturers to introduce less toxic products. "We hope we are transforming the marketplace by doing this," said Sierra Club executive director Carl Pope. "These products are clean, they're green, they're not going to hurt you, and they're not going to hurt the environment." See Ilana DeBare, "Clorox Introduces Green Line of Cleaning Products," *San Francisco Chronicle*, January 14, 2008.

135 **Martin Wolf is unmoved:** "We're committed to having a science-based approach to sustainability," said Lauren Thaman, a P&G public affairs specialist. "We have taken a life-cycle analysis approach to ensure we are making claims that are meaningful and measureable. I will tell you that until being able to say something is 'natural' means it is more sustainable, we will not do that." See Michael McCoy, "The Greening Game," *Chemical & Engineering News*, January 26, 2009.

136 **To Martin Wolf:** Marcelle S. Fischler, "A Safe House?," *New York Times*, February 15, 2007, and Howard Fine, "Dry Cleaning Blues," *Los Angeles Business Journal*, February 4, 2008.
In the last couple of years: Jeffrey Hollender, "Has Seventh Generation Sold Out by Working with Wal-Mart?" GreenBiz.com, October 29, 2008; www.greenbiz.com/blog/2008/10/29/strange-bedfellows-seventh-generation-wal-mart?p.

CHAPTER FOUR: THE TAP

141 **Although thousands of synthetic toxins:** Charles Duhigg, "That Tap Water Is Legal but May Be Unhealthy," *New York Times*, December 16, 2009.
"Surprisingly little is known": Dana W. Kolpin et al., "Pharmaceuticals, Hormones, and Other Organic Wastewater Contaminants in U.S. Streams, 1999-2000: A National Reconnaissance," *Environmental Science and Technology*, vol. 36, no. 6 (2002), pp. 1202–11.
Some 97 percent: U.S. Geological Survey, "Pesticides in the Nation's Streams and Ground Water, 1992–2001: A Summary," *Fact Sheet 2006–3028, March 2006*. The report can be found at water.usgs.gov/nawqa. See also Charles Duhigg, "Debating How Much Weed Killer Is Safe in Your Water Glass," *New York Times*, August 22, 2009; Theo Colburn, Dianne Dumanoski, and John Peterson Myers, *Our Stolen Future* (New York: Plume Books, 1996); and www.pollutioninpeople.org/results/report/chapter-6/ddt_pcb_1.

144 **Kauffman spends a good bit:** Dawn Fallik, "Drinking Water Holds Surprises," *Philadelphia Inquirer*, November 28, 2004.

145 **"Water treatment systems":** After surveying mountains of data from state water departments in 2005, the Environmental Working Group reported that 260 contaminants were routinely found in the

nation's tap water, including 141 that don't even have enforceable safety limits. Dozens of these unregulated chemicals have been linked to cancer, reproductive problems, and immune system damage. More distressing, the EWG found that states spend just a tiny fraction of the billions of dollars they receive through the federal Clean Water Act on mitigating pollution running off farms and city and suburban streets, which makes up 60 percent of water pollution. By not controlling pollution at its source—by not convincing farms and shopping malls to control runoff—states instead depend on massive treatment plants to clean water up at the back end. This, by anyone's estimation, is a poor trade-off. "We find a deep disconnect between what people care about and what the government is willing to act upon," the EWG report concluded. "From agricultural pollution, to industrial waste, to pollution stemming from sprawl and urban runoff, a lack of political will materializes into poor planning and scarce funding that leads to pollution beginning upstream and ending at the tap." See Environmental Working Group, "A National Assessment of Tap Water Quality," December 20, 2005; www.ewg.org/tapwater/findings.php.

145 **Before the federal Clean Water Act:** Charles Duhigg, "Clean Water Laws Are Neglected, at a Cost in Suffering," *New York Times*, September 12, 2009.

146 **To be fair:** See the EPA website at www.epa.gov/reg3wapd/tmdl/pa _tmdl/ChristinaMeetingTMDL/NutrientLDO/Sec2-CR_Nutrient _TMDL-20060118.pdf.

147 **A few years ago:** Jennifer Roberts, Ellen Silbergeld, and Thaddeus Graczyk, "A Probabilistic Risk Assessment of Cryptosporidium Exposure Among Baltimore Urban Anglers," *Journal of Toxicology and Environmental Health, Part A*, vol. 70 (2007), pp. 1568–76.
Now, you might say: CNN, September 2, 1996.
In the spring of 2008: Ellen Silbergeld and Jay Graham, "The Cuyahoga Is Still Burning," *Environmental Health Perspectives*, vol. 116, no. 4 (April 2008). Waterborne pathogens are not included in the fish advisories mandated by the Clean Water Act of 1977. And despite the growing prevalence of things like *E. coli* and cryptosporidium, the way pathogens are measured has not been updated since the 1960s.

148 **The problem is not just animal waste:** See the EPA website at

cfpub.epa.gov/npdes/home.cfm?program_id=4. See also Elizabeth
Royte, "A Tall, Cool Drink of . . . Sewage?," *New York Times
Magazine*, August 10, 2008.

148 **In Baltimore, sewage treatment systems:** www.justice.gov/opa/pr/
2005/July/05_enrd_387.htm.

"It's hard to quantify": Charles Duhigg, "Clean Water Laws Are
Neglected, at a Cost in Suffering," *New York Times*, September 12,
2009.

And what is true in miniature: Timothy Wheeler, "Chicken Farmers
Face Strict EPA Rules," *Baltimore Sun*, March 15, 2009.

149 **But given American history:** Felicity Barringer, "Reach of Clean
Water Act Is at Issue in 2 Supreme Court Cases," *New York Times*,
February 20, 2006. See also "A Clear, Clean Water Act," *New York
Times*, April 16, 2009, and Gerald Kauffman, "What If . . . the
United States of America Were Based on Watersheds?," *Water Policy*,
vol. 4 (2002), pp. 57–68.

150 **Just a couple of weeks before:** Jeff Donn, Martha Mendoza, and
Justin Pritchard, "An AP Investigation: Pharmaceuticals Found in
Drinking Water," Associated Press, March 10, 2008.

153 **Another enormous headache:** PBS's *Frontline*, "Poisoned Waters,"
April 21, 2009. See also Timothy Wheeler, "Chicken Growers Face
EPA Crackdown," *Baltimore Sun*, March 15, 2009.

159 **Fluoride has been added:** See the congressional testimony of Dr.
William Hirzy, National Treasury Employees Union Chapter 280,
June 29, 2000, at www.fluoridealert.org/testimony.htm.

In 2001, the CDC: Sharon Begley, "Government Panel Raises
Concern About Fluoride," *Wall Street Journal*, March 23, 2006. See
also Margot Roosevelt, "Not in My Water Supply," *Time*, October 17,
2005.

161 **The story doesn't end here:** Thomas Webler et al., "Exposure to
Tetrachloroethylene via Contaminated Drinking Water Pipes in
Massachusetts: A Predictive Model," *Archives of Environmental
Health*, vol. 48 (1993), pp. 293–97. See also Ann Aschengrau et al.,
"Cancer Risk and Tetrachloroethylene-Contaminated Drinking
Water in Massachusetts," *Archives of Environmental Health*, vol. 48
(1993), pp. 284–92, and Charles D. Larson et al.,
"Tetrachloroethylene Leached from Lined Asbestos-Cement Pipe
into Drinking Water," *Research and Technology*, April 1983.

161 **In the Maine body burden study:** Ellen Silbergeld, testimony before the U.S. House of Representatives Committee on Government Reform, Hearing on Lead in D.C. Water and Sewer Authority Water, in *Public Confidence Down the Drain: The Federal Role in Ensuring Safe Drinking Water in the District of Columbia* (GPO, March 5, 2004). Even as contamination from one anti-knock fuel additive diminishes, others arise, Jerry Kauffman says. "We've seen some fantastic dips in environmental lead," Kauffman said. "The trouble is, now we're seeing all this MBTE ending up in the groundwater."

162 **Since some bottlers:** Elizabeth Royte, "A Fountain on Every Corner," *New York Times*, May 23, 2008.

CHAPTER FIVE: THE LAWN

168 **My students had never heard:** The other constituent is 2,4,5-T (2,4,5-trichlorophenoxyacetic acid), which has since been outlawed because of its links to cancer, miscarriages, and birth defects.

Back in our little circle: The war ended before field tests could be completed, but—as happened with so many other chemicals developed during the war—2,4-D quickly proved useful in another realm, in this case for killing broad-leaved plants.

In the 1940s: Gale Peterson, "The Discovery and Development of 2,4-D," *Agricultural History*, vol. 41, no. 3 (July 1967), pp. 243–54. So why do we keep using the stuff? It can't be the science. Between 1944 (when 2,4-D was developed) and 2002, more than 750 scientific papers were published on the best ways to control weeds with chemicals. In those same 58 years, guess how many papers were published looking at achieving the same results with things like better mowing techniques, more effective fertilizers, and smarter grass seed? Just 25. See Ted Steinberg, *American Green: The Obsessive Quest for the Perfect Lawn* (New York: Norton, 2006), p. 219. While most of the health problems associated with Agent Orange have been attributed to dioxin contamination of the compound's other component, 2,4,5-T, which has since been banned, several forms of dioxin have also been found in 2,4-D.

169 **Here's why: beyond its ability:** Paul Robbins and Julie Sharp, "The Lawn-Chemical Economy and Its Discontents," *Antipode*, vol. 35 (November 2003), pp. 955–79. In the summer of 1944, researchers

sprayed 2,4-D on a dandelion-infested lawn at the federal Plant
Industry Station in Beltsville, Maryland. They achieved "a complete
kill." When this news reached members of the U.S. Golf Association,
they immediately tested the chemical on a golf course. It killed 80
percent of the clover, but left the bluegrass intact. How it did this
was something of a mystery.

169 **At first, 2,4-D's impact:** To say that 2,4-D is everywhere is not an ex-
aggeration. In the most comprehensive analysis of its kind, the U.S.
Geological Survey reported in 2006 that pesticides were found in
every one of the 139 American streams it studied. One-fifth of the
streams had ten or more. And the USGS was looking only for pesticides; it wasn't looking for flame retardants, or phthalates, or anything else. What it found was newsworthy enough, including the fact
that compounds that have not been used in decades, like DDT, were
found both in fish and in the streambeds of most of the rivers—even
in rivers in "undeveloped" countryside. A handful of pesticides, including 2,4-D, showed up more frequently and at higher concentrations in city streams than in nonurban ones, in part because so many
chemicals were running off farms and suburban lawns upstream. See
U.S. Geological Survey, "Pesticides in the Nation's Streams and
Groundwater, 1992–2001: A Summary"; pubs.usgs.gov/fs/2006/
3028/pdf/fs2006-3028.pdf.

Given its effect: S. K. Hoar et al., "Agricultrual Herbicide Use and a
Risk of Lymphoma and Soft-Tissue Sarcoma," *Journal of the
American Medical Association*, vol. 256, no. 9 (1986), pp. 1141–47.
Other information on 2,4-D is from *ChemicalWatch Factsheet*, avail-
able at www.beyondpesticides.org.

170 **Five years later:** H. M. Hayes et al., "Case-Control Study of Canine
Malignant Lymphoma: Positive Association with Dog Owner's Use
of 2,4-D Herbicides," *Journal of the National Cancer Institute*, vol. 83,
no. 17 (1991), pp. 1226–31, and Marcia G. Nishioka et al.,
"Measuring Transport of Lawn-Applied Herbicide Acids from Turf
to Home: Correlation of Dislodgeable 2,4-D Turf Residues with
Carpet Dust and Carpet Surface Residues," *Environmental Science
and Technology*, vol. 30, no. 11 (1996), pp. 3313–20. "Transport of
pesticides into the home carries significant implications for human
exposure," another study reports. "Carpets, house dust, and home
furnishings become long-term sinks for the pesticides; the common

environmental weathering factors such as wind, rain, soil microbes, and sunlight are not available for degradation. Residues on floors and surfaces can become a source of exposure for young children through the hand to mouth route of ingestion, as has been fully documented for lead exposure." See Ruthann Rudel et al., "Phthalates, Alkylphenols, Pesticides, Polybrominated Diphenyl Ethers, and Other Endocrine-Disrupting Compounds in Indoor Air and Dust," *Environmental Science and Technology*, vol. 37, no. 20 (2003), pp. 4543–53, and Marcia G. Nishioka et al., "Distribution of 2,4-D in Air and on Surfaces Inside Residences After Lawn Applications: Comparing Exposure Estimates from Various Media for Young Children," *Environmental Health Perspectives*, vol. 109, no. 11 (November, 2001), pp. 1185–91.

170 **Of course, 2,4-D is just one:** David Pimentel, "Silent Spring Revisited," *Journal of the Royal Society of Chemistry*, October 2002.

171 **A hundred years ago:** Elizabeth Kolbert, "Turf War," *New Yorker*, July 21, 2008.

I'm not sure if 50 million acres: Douglas Tallamy, *Bringing Nature Home: How Native Plants Sustain Wildlife in Our Gardens* (Portland, Oreg.: Timber Press, 2007), pp. 117–88. See also Ted Steinberg, *American Green*, p. 4. The gasoline spill numbers can be found at www.epa.gov/glnpo/greenacres/toolkit/chap2.html.

172 **And like most synthetic things:** Ted Steinberg, *American Green*, pp. 19–24. Abe Levitt's piece "Chats on Gardening," printed in the Levittown *Times* on June 17, 1948, is reported in *American Green*, p. 24.

173 **"Faster than a garbage can":** "The Wicked Weed," *Time*, September 7, 1959.

Connecting (and encouraging): Rachel Carson, *Silent Spring* (Boston: Houghton Mifflin), 1962, pp. 69–75. Carson wrote acidly of the "weed killer's philosophy" she once found in the proceedings from a weed-control conference. The author "defended the killing of good plants 'simply because they are in bad company.' Those who complain about killing wildflowers along roadsides reminded him, he said, of antivivisectionists 'to who, if one were to judge by their actions, the life of a stray dog is more sacred that the lives of children.' To the author of this paper, many of us would unquestionably be suspect, convicted of some deep perversion of character because we prefer the sight of the vetch and the clover and the wood lily in all their delicate and transient beauty to that of roadsides scorched as by fire, the shrubs brown and brittle, the bracken that once lifted high its brown lacework now withered and drooping. We would seem deplorably weak that we can tolerate the sight of such 'weeds,' that we do not rejoice in their eradication, that we are not filled with exultation that man has once more triumphed over miscreant nature." Carson noted that of the 70 species of shrubs and vines commonly found on eastern roadsides, 65 are important sources of food for wildlife, including many species of bees and other insect pollinators, which depend on "weeds" like goldenrod, mustard, and dandelions for pollen and food for their young.

174 **In 1968, a former garden store owner:** In 1992, ChemLawn was purchased by ServiceMaster, a conglomerate founded in 1929 as a Chicago moth-proofing company, and folded into a company called TruGreen. Today, TruGreen is the largest lawn and landscaping company in the world, with 3.5 million customers. ServiceMaster also owns Merry Maids, Terminix, and Rescue Rooter. The ServiceMaster website reports that the company's founder, Marion E. Wade, "had a strong personal faith and a desire to honor God in all he did. Translating this into the marketplace, he viewed each individual employee and customer as being made in God's image—worthy of dignity and respect." Eighty years later, the company's objectives remain in place: "To honor God in all we do; To help people develop; To pursue excellence; and To grow profitably." See P. Ranganath Nayak and John M. Ketteringham, *Breakthroughs!* (New York: Rawson Associates, 1986), pp. 74–81; the ServiceMaster website can be found at corporate.servicemaster.com/overview_history.asp.

175 **By 1999, more than two-thirds:** Paul Robbins and Julie Sharp, "The Lawn-Chemical Economy and Its Discontents," *Antipode*, vol. 35 (November 2003), pp. 955–79.

Lawn-care companies: When First Lady Michelle Obama announced in the spring of 2009 that she wanted an organic vegetable garden installed within the White House lawn, a group called the Mid America CropLife Association, representing agribusinesses like Monsanto, Dow AgroSciences, and DuPont Crop Protection, warned that growing vegetables without the use of chemicals was impractical, even elitist. "Americans are juggling jobs with the needs of children and aging parents," the industry group said. "The time needed to tend a garden is not there for the majority of our citizens, certainly not a garden of sufficient productivity to supply much of a family's year-round food needs." Garden chemicals, once a shield against a home owner's being perceived as a Communist, can now be thought of as a time-saving Family Value. John Nichols, "Manure!," *Nation*, May 4, 2009.

178 **River-protection groups:** See Chesapeake Bay Foundation, "The State of the Bay, 2007."

   In California, scientists are discovering: "California Sea Lions Seizures May Come from Fetal Domoic Acid Poisoning," National Oceanic and Atmospheric Administration, June 9, 2008.

179 **And it's not like Americans:** John Grossman, "How Green Are Those Fairways?," *Audubon*, September–October 1993.

   Which brings me: C. Leonard et al., "'Golf Ball Liver'": Agent Orange Hepatitis," *Gut*, vol. 40 (1997), pp. 687–88. See also J. Johnston et al., "'Golf Ball Liver'": A Cause of Chronic Hepatitis?," *Gut*, vol. 42 (1998), pp. 143–46.

181 **"I didn't mince words":** Churchill had come to his own wisdom circuitously. Trained as an agricultural scientist in the 1960s, he was not at first opposed to using synthetic chemicals to control plant growth. Reading *Silent Spring* had opened his eyes, but it was a trip to the agricultural ministry in India ten years later that changed his mind. How was it, Indian scientists had asked him, that the United States could ban DDT in its own country but still sell it to developing countries like India? They accused Churchill of "being a party to the poisoning of the people of their country." Churchill vowed to change his ways.

182 **Tukey started doing some research:** Jack K. Leiss and David A. Savitz, "Home Pesticide Use and Childhood Cancer: A Case-Control Study," *American Journal of Public Health*, vol. 85, no. 2 (February 1985), pp. 249–52.

   Tukey also learned that exposure: "Clearly, the acute effects of pesticides are well known," Dr. Aaron Blair of the National Cancer Institute has said. "After all, they are toxic chemicals, and they are constructed to be that way. People are poisoned by them all the time." Sandy Rovner, "To Spray or Not to Spray," *Washington Post*, July 7, 1987; see also V. F. Garry, "Pesticide Appliers, Biocides, and

Birth Defects in Rural Minnesota," *Environmental Health Perspectives*, vol. 104 (1996), pp. 394–99. Domestic pesticides are regulated by the Federal Insecticide, Fungicide, and Rodenticide Act, known as FIFRA, which requires the EPA to evaluate the risks and benefits of chemicals before they are placed on the market—and also to go after companies making dubious health claims about their chemical products. There have been some odd lapses in this law over the years: diazinon, a pesticide used widely for decades (and with a similar chemical composition to nerve gas) was banned (in phases) by the EPA beginning in 2000, but retailers were allowed to continue selling it until 2004.

Oddly, the law does not apply to "professional applicators" like lawn-care companies; in a kind of Catch-22, these companies are regulated by the Federal Trade Commission, which says it does not enforce the laws because it prefers to defer to the EPA "in such matters because of EPA's expertise and legislative authority." In other words, the EPA has no control over lawn-care companies, and the FTC, which does, does nothing in deference to the EPA, which does not. And so it goes.

184 **Quitting an addiction:** At the time, Americans were spraying 70 million pounds of pesticides on their lawns and golf courses, all of them, presumably, registered with the EPA. But EPA registration is not a determination of safety, said Phyllis Spaeth, an assistant state attorney general in New York: "It's a balancing act, a cost-benefit analysis. If there is a chemical need for agriculture, for example, and it's the cheapest one, then the EPA says, 'Well, it's the only one that's out there, so we'll let them use it.'" ChemLawn refused to admit wrongdoing but settled the lawsuit for $100,000 and agreed not to produce ads making health claims. The ChemLawn settlement was reached around the time the Government Accountability Office released a report critical of the claims made by lawn-care companies. The report noted that regulation of lawn services fell in the gray area between the EPA and the Federal Trade Commission. A spokesman for the EPA said any lawn-care company making safety claims was "perpetrating a hoax." Pesticides are toxic, he said, "by their very nature."

The GAO report prompted some lawn-care companies to rein in their health claims. Lawn Doctor, a New Jersey company, stopped claiming that its products were "practically nontoxic to humans, pets and the environment." Lawn Doctor's promotional materials now say that the products are considered safe when they are applied according to instructions on the labels that have been approved by the federal government. "We don't want to mislead the public," a company spokesman said. "We have no intention of doing that."

In January 1994, under similar pressure, DowElanco agreed to pull its advertisements claiming that the herbicides Garlon and Tordon are less toxic than aspirin and caffeine, and that the insecticide Dursban LO shows "no evidence of significant risk to the environment"—despite a warning on its own label that the poison "is toxic to birds and other wildlife and extremely toxic to fish and aquatic organisms."

A few years later, in 1996, New York's attorney general ordered Monsanto to pull ads that said its herbicide Roundup was "safer than table salt" and "practically nontoxic" to mammals, birds, and fish. The company withdrew the spots, but also said that the phrase in question was permissible under EPA guidelines. A year later, Monsanto recruited Cary Sharp, the horticulturist for the San Diego Zoo, to pitch Roundup on television. In the ad, Sharp described his use of Roundup at work and at home. "The sun brings life. People come to the park to see animals, not weeds," Sharp says in the advertisement. As he talks, the camera pans across lush landscapes of exotic plants and animals. "Pulling weeds is a waste of time," he remarks. "We use Roundup. Roundup kills the whole weed, roots included." He ends the spot with the tagline "Roundup. No roots. No weeds. No problem."

The $14 million ad campaign ran in forty-eight local markets as well as on national broadcast and cable networks. Monsanto also promoted its "Roundup for Species Survival" program, through which it donated a percentage of Roundup sales to local zoos and endangered-species programs. "These zoos are looking for auxiliary income," Mike McGrath, the editor of *Organic Gardening* magazine, told the *New York Times.* "They want partnerships, and they may not care that these are chemical companies. Or maybe they believe what the chemical companies say." See Tamar Charry, "Monsanto Recruits the Horticulturist of the San Diego Zoo to Pitch Its Popular Herbicide," *New York Times*, May 29, 1997. See also Barry Meier,

"Lawn Care Concern Says It Will Limit Safety Claims," *New York Times*, June 30, 1990; Anne Raver, "Fertilizing Your Lawn? Look Before You Leap," *New York Times*, April 12, 1994; and John Scow, "Can Lawns Be Justified?," *Time*, June 3, 1991.

185 **In 2003, TruGreen ChemLawn:** A coalition of child advocacy and environmental groups pressured U.S. Youth Soccer to terminate the sponsorship. In a letter (complete with footnotes listing scientific studies), the groups asked Youth Soccer to cut its ties with the lawn-care company. "Clearly, TruGreen/ChemLawn wants to enlist children as allies to ask or nag their parents for lawn services so that any parent who chooses not to hire ChemLawn will be viewed by their children as not caring about youth soccer," the letter said. "To put it simply, TruGreen/ChemLawn marketers want to make it as hard as possible for parents to say no."

Why should parents reject the offer from ChemLawn? "Like so much that is marketed to or through children, the letter noted, ChemLawn's pesticides can be harmful to them. Studies have linked lawn pesticides to birth defects, liver and kidney damage, and neurological disorders. Young children and adolescents are particularly susceptible to the toxic effects of lawn chemicals, [and] the use of pesticides has been linked to an increased risk of childhood illnesses including non-Hodgkin's lymphoma, brain cancer, and leukemia. Of the three active ingredients in ChemLawn's popular Tri-Power system, two (MCPA and Mecoprop-P) are possibly carcinogenic; the third (Dicamba) is classified by the Pesticide Action Network as a developmental and reproductive toxin."

The campaign apparently worked; TruGreen ChemLawn pulled its sponsorship in 2004. See Erica Noonan, "Environmental Group Targets Developer," *Boston Globe*, December 7, 2003.

A study conducted by University of Texas pediatricians looked at 20 children who had been referred to them by other hospitals and whom they properly diagnosed as victims of pesticide poisoning. They found that 16 of the 20 had been misdiagnosed before the referral. "Initial diagnoses included pneumonia, bronchitis, diabetes, brain aneurysm, and head trauma," the study showed. "In each of those cases, the symptoms were actually caused by exposure to organophosphate or carbamate pesticides. Both of these types of pesticides are among those frequently selected for use by our survey respondents." See Michael H. Surgan, Thomas Congdon et al., "Pest Control in Public Housing, Schools and Parks: Urban Children at Risk," Environmental Protection Bureau, Office of the New York State Attorney General, August 2002. The full report can be found at www.oag.state.ny.us/bureaus/environmental/pdfs/pest_control _public_housing.pdf.

185 **Even Long Island:** Bruce Lambert, "A Verdant Lawn as a Health Issue: Nassau Bill Would Require Notice Before Pesticide Spraying," *New York Times*, March 22, 1996. See also Long Island Neighborhood Network, www.longislandnn.org/pesticides/alert _nn2.htm#history.

186 **It's not just Canada:** "Euro MPs Back Pesticide Controls," BBC News, January 13, 2009; "EU Pesticides Will 'Wipe Out' Carrot Crop," *Guardian*, January 4, 2009; "New EU Pesticides Law Falls Short of Real Progress," Greenpeace European Unit, January 13, 2009.

187 **The group's lobbyists:** So far, their efforts appear to have been successful. Forty-one states have these preemptive laws on the books. A California bill returning control of pesticide spraying to towns was killed after intense lobbying.

Yet there are signs of resistance. Madison, Wisconsin, banned the use of phosphorous in lawn fertilizer in 2004; one study determined that Wisconsin home owners already had enough phosphorous on their lawns for the next eight to twenty-two years if they would just leave their grass clippings on the lawn when they mowed. That same year, the seven counties that make up Minnesota's Twin Cities banned phosphorous, and a year later, the chemical was banned in the entire state of Minnesota.

In the last few years, Westchester County, New York, which maintains 18,000 acres of public space, has effectively eliminated the use of pesticides in its parks and greatly reduced their use on golf courses. (There are 5,000 acres in the county under corporate stewardship that are not under county landscaping control.) The county has also become partners with the Grassroots Environmental Education Program, which has trained hundreds of private landscapers in nonchemical methods, and the insurance company Swiss Re's United States headquarters in Armonk joined the program, eliminating pesticide use on its 127-acre corporate campus. "We are blessed with great water here," the county executive told the *New York Times* in 2007. "There aren't many places in the world that have such good water, so for us to contaminate it is really criminal. These pesticides get into the water, and they are unhealthy for people and animals." See Abby Gruen, "Homeowners Seek Safer Alternatives to Pesticides," *New York Times*, April 1, 2007.

In April 2009, just four months after the departure of the Bush administration, the EPA said it would order manufacturers of 67 pesticides to test whether their products disrupt hormonal systems in people or animals. The chemicals were picked not because of their acute toxicity but because of their widespread use, and the EPA plans to order that hundreds more chemicals be tested in the coming years. I wouldn't recommend holding your breath; Congress passed a law mandating such tests in 1996. Industry is confident that the chances of problems appearing are "extremely low," said Jay Vroom, president and chief executive of CropLife America, a major trade association and an affiiate of the lobbying group RISE. Vroom said the pesticide industry was "very confident our products will come through with flying colors." See Juliet Eilperin, "EPA Will Mandate Tests on Pesticide Chemicals," *Washington Post*, April 16, 2009.

The Government Accountability Office issued a report in 1990 that said the EPA had made little progress bringing some 24,000 pesticide products into compliance. Worse, despite a pesticide industry that made safety claims the EPA considered to be "false and misleading," the EPA had done little to crack down on them. The lawn pesticides industry "continues to make prohibited claims that its products are safe or nontoxic," the GAO reported. For the two most frequently used pesticides, diazinon and 2,4-D, the EPA "identified possible health effects associated with their use" but did not then push for further assessment or regulation. Until such work is done, "the public's health may be at risk from exposure to these pesticides," the GAO reported. (The EPA banned diazinon from use on golf courses and sod farms in 1988, after learning that the chemical had been killing geese. But it still allowed it to be sold for residential use until 2004, and it is still legal to use diazinon purchased before then. Tom Adamczyk, an EPA herbicide official, claimed that geese do not typically land on suburban lawns, and that in any case the

EPA did not have the money or the personnel to act more swiftly once chemicals are proven harmful. "You can't just yank a product off the market without incontrovertible proof that it's harmful," he said.) See General Accounting Office, "Lawn Care Pesticides: Risks Remain Uncertain While Prohibited Safety Claims Continue," March 1990. The full report can be found at http://161.203.16.4/d24t8/140991.pdf.

190 **"In an act":** Doug Tallamy, *Bringing Nature Home*, p. 22.

191 **So here's a question:** Y. Wang, C. Jaw, and Y. Chen, "Accumulation of 2,4-D and Glyphosate in Fish and Water," *Water Air Soil Pollution*, vol. 74 (1994), pp. 397–403; B. L. Robers and H. W. Dorough, "Relative Toxicity of Chemicals to the Earthworm," *Environmental Toxicology and Chemistry*, vol. 3 (1984), pp. 67–78; Caroline Cox, "Herbicide Factsheet: 2,4-D; Ecological Effects," *Journal of Pesticide Reform*, vol. 19, no. 3 (1999), pp 14–19.

The toll of suburban development: David Pimentel, "Silent Spring Revisited," *Journal of the Royal Society of Chemistry*, October 2002.

192 **It's not just lawn chemicals:** Doug Tallamy, lecture given at Maryland's Irvine Nature Center, August 24, 2008; Bridget Stutchbury, "Did Your Shopping List Kill a Songbird?," *New York Times*, March 30, 2008; www.stateofthebirds.org.

In the last year: Major efforts to restore wetlands are helping restore populations of herons, egrets, and ducks in some areas; John Hoskins, of the United States North American Bird Conservation Initiative, an umbrella group of public and private conservation efforts, says he is hopeful. "When agencies, organizations, and individual citizens work together to conserve precious resources, some really good things happen." See Cornelia Dean, "Nearly a Third of Bird Species in U.S. Are Found Declining," *New York Times*, March 20, 2008, and Tallamy, *Bringing Nature Home*, p. 30.

"It is curious that the news media": Tallamy, *Bringing Nature Home*, pp. 25–28.

193 **Tallamy's vision is based:** Tallamy, *Bringing Nature Home*, pp. 83–85.

194 **I asked Tallamy:** The notion of invasive species—and what to do about them—is a complicated one. Isn't the United States, after all, known as a country of immigrants? Wouldn't adding plants and animals from faraway places similarly enrich our lives? The trouble is, unlike the diversity created by immigrant people, the "diversity" cre-

ated by invasive plants does not enhance the native population. Take the paperbark melaleuca, an ornamental tree imported from Australia. In Florida, the paperbark has spread so aggressively in the Everglades that it has now covered hundreds of thousands of acres, and the resulting forests offer nothing to native insects, birds, or reptiles. Alligators and birds don't nest there, and since native insects don't feed on paperbark leaves, there is nothing for birds to eat. The addition of a single species, Tallamy notes, has caused an entire ecosystem to collapse.

Or take the Norway maple. Introduced to Philadelphia's Morris Arboretum 250 years ago, it has spread all over the eastern seaboard. Yet for insects, and thus for birds, it is still virtually useless as a source of food. Why? Because in the 80 million years that have elapsed since the European and North American continents separated, the Norway maple has been evolving apart from North American insects. In the context of 80 million years, Tallamy says, 300 years means nothing.

Or take a plant that has become the very symbol of an East Coast spring: the azalea. Although there are plenty of native azaleas to choose from, all of which support native insects (and thus native birds), gardeners still flock to buy exotic species from Japan, which support exactly one insect: the azalea lace bug, and none of its natural predators. When lace bug populations explode, gardeners reach for their spray cans. And around and around we go.

Or take bees. There has been much in the news in the last few years about the collapse of bee colonies, a frightening development that still has scientists scratching their heads. To Tallamy, the problem has fairly simple roots. Just as agribusiness has turned cattle into commodities, forcing grass eaters to become corn eaters, so has it forced bees to survive on single sources of pollen.

"We truck bees to California from Pennsylvania, from Texas, from Georgia, and make them pollinate almonds," he said. "But almond trees offer only one source of pollen. Everyone knows a single source of pollen is a recipe for trouble. These are very dry regions. There are no flowering plants out there. Since 'good farming practices' means eliminating all weeds, farmers have essentially eliminated all other sources of food."

194 **A couple of years ago:** "Nurserymen are not evil people who

planned to eliminate the American chestnut in the East with chestnut blight, or sugar pines in the West with blister rust, or Fraser firs
in the Smokies with balsam wooly adelgids, or hardwood and fruit
trees in the Pacific Northwest with citrus long-horn beetles or
American beech with beech scale, or the citrus industry in Florida
with Asian psyllids, or oaks throughout the country with sudden oak
death disease," Tallamy writes. "The introduction of these organisms,
particularly those that occurred in the early 1900s, was an accident
that happened because we had little understanding of the consequences of introducing foreign organisms with which our native
plants and animals had no evolutionary experience. I am more critical of our behavior today, because we do know what can happen
when we import alien plants. It has happened over and over again.
Could it be that both the gardening public and the nursery industry
consider the elimination of key species from entire ecosystems to be
'collateral damage,' in the parlance of the military, an undesirable
but unavoidable consequence of creating beautiful gardens with de-
sirable exotic ornamentals?" See Tallamy, *Bringing Nature Home*,
p. 69.

194 **Native trees, Tallamy has since found:** Tallamy, *Bringing Nature
Home*, p. 21.

197 **On the other side:** Like most evangelicals, indeed, like SafeLawns'
Paul Tukey, Doug Tallamy speaks with the passion of the convert.
Some years after his eyes were opened by the bulldozer, Tallamy
went off to graduate school at the University of Maryland, where he
found the mysteries of exotic plants so enthralling that he would
germinate seeds in the university greenhouse, then rush off and
plant them all over his parents' property: Japanese hardy oranges
from Japan; royal paulownia trees from China. The trees were beautiful, and best of all, none of his neighbors had them.

At the same time, Tallamy was enrolled in graduate courses that
taught him that herbivorous insects can survive only by eating plants
they have evolved alongside for millions of years. To an untrained
human eye, a leaf is a leaf is a leaf—why can't insects just pick and
choose? In the laboratory, Tallamy learned that insects see—and
taste—things differently. If you took a pair of leaves from a white
oak tree and an ailanthus (the Chinese tree of heaven) and ground
them up, you'd find they had entirely different chemical con-

stituents. Each compound gives its leaf a peculiar taste; to some insects it might be a delicacy, to others, having evolved differently, it might be a toxin. Ninety percent of leaf-eating insects can eat three plant types or fewer. But each species of insect has figured out a way, over countless generations, to digest the leaves of specific plants. More interestingly, some of the leaves that insects eat actually make the insects themselves toxic to birds. Milkweed, for example, makes monarch butterfly caterpillars so distasteful to birds that the gaudy yellow-and-black caterpillars are left to munch away undisturbed.

Worldwide, 37 percent of animal species are herbivorous insects. They convert the tissues of plants into insect tissue, and thus food for other species. In North America, 96 percent of terrestrial bird species rely on insects and arthropods (mostly spiders that eat insects) for food. Pound for pound, insects contain more protein than beef. "And because it is we who decide what plants will grow in our gardens, the responsibility for our nation's boidiversity lies largely with us," Tallamy writes. "We have proceeded with garden design as we always have, with no knowledge of the new role our gardens play—and alas, it shows. Throughout suburbia, we have decimated the native plant diversity that historically supported our favorite birds and animals." See Tallamy, *Bringing Nature Home*, pp. 104–5.

EPILOGUE

204 **In 1989, the NRDC:** Doug Haddiz, "Alar as Media Event," *Columbia Journalism Review*, March–April 1990.

206 **In a column in the magazine's:** *Green Guide*, Spring 2008.

207 **"Of all the chemical products":** Stacey Malkan, *Not Just a Pretty Face: The Ugly Side of the Beauty Industry* (Gabriola Island, B.C.: New Society Publishers, 2007), p. 137.

208 **In recent years:** Milt Freudenheim, "Maker of IV System to Stop Using a Plastic," *New York Times*, April 7, 1999. See also Susan Moran, "A Turn to Alternative Chemicals," *New York Times*, March 26, 2008.

Since 1996, the EPA: William McDonough and Michael Braungart, *Cradle to Cradle: Remaking the Way We Make Things* (New York: North Point Press, 2002); Deborah Solomon, "Calling Mr. Green," *New York Times*, May 20, 2007.

209 **Last year, in an effort:** Susan Moran, "Panning E-Waste for Gold," *New York Times*, May 17, 2006.

DuPont is working: Yudhijit Bhattacharjee, "Harnessing Biology, and Avoiding Oil, for Chemical Goods," *New York Times*, April 9, 2008.

212 **"Hey, Ladies, I am participating":** Pingree's email was included in a speech she gave at the Common Ground Fair, September 22, 2007.

213 **For starters, it grandfathered in:** The EDF study can be found at www.environmentaldefense.org/documents/243_toxicignorance .pdf.

For another thing: See GAO-05-458, the United States Government Accountability Office, "Chemical Regulation: Options Exist to Improve EPA's Ability to Assess Health Risks and Manage Its Chemical Review Program," June 13, 2005. TSCA was supposed to require companies to give ninety days' notice before releasing a new substance onto the market, to allow the EPA to assess a chemical's toxicity and potential for public exposure. The law did not say that companies had to generate toxicity data, so most companies didn't. And if ninety days seems like an inadequate amount of time to make such an important call, consider this: over the last thirty years, EPA's reviews have resulted in action being taken to reduce risks of just 11 percent of the 32,000 newly developed chemicals. More troubling, TSCA does not require companies to test new chemicals for toxicity or to gauge exposure levels, and most companies don't. In the late 1990s, the EPA began something called the Voluntary Children's Chemical Evaluation Program, which began by identifying 23 chemicals—of the more than 80,000 now on the market—and asking manufacturers to disclose what they knew about the chemicals' impact on children's health. Participation was not mandatory. Companies that did participate were asked to submit their findings to a panel of scientists, which would then declare whether the chemicals were safe for use around children. Chemical industry lobbyists called the program "breathtakingly significant." But the program lacked credibility from the beginning. Companies regularly refused to submit what they knew or ignored requests from the EPA. In one case, the EPA asked 12 chemical companies for tests to assess the safety of benzene, a chemical used in gasoline that is known to cause leukemia, anemia, and bone marrow disease. The EPA wanted information about how the chemical can affect the developing brain and reproductive system. The chemical companies, including BP Amoco, Dow Chemical, and ExxonMobil Chemical, said they disagreed with the government's opinion that more information was needed and refused to provide the answers. In the case of xylenes, a chemical found in gasoline, paint varnish, shellac, and cigarette smoke, the companies simply did not reply when the EPA requested further testing. Some oversight panels were stacked with scientists with ties to the chemical industry. Ten years later, the program is basically defunct. See Susanne Rust and Meg Kissinger, "EPA Drops Ball on Danger of Chemicals to Children," *Milwaukee Journal Sentinel*, March 30, 2008.

214 **Today, fully 85 percent:** The EPA's Office of Pollution Prevention, which administers TSCA, is widely criticized as being too close to industry. But even if the office wanted to do its work, its budget is perpetually underfunded. Everyone from the General Accounting Office to the National Academy of Sciences to the Congressional Office of Technology Assessment to the EPA itself has identified areas where TSCA has fallen short, but it has not been significantly changed in thirty years. "EPA's reviews of new chemicals provide only limited assurance that health and environmental risks are identified because TSCA does not require companies to test chemicals before they notify EPA of their intent to manufacture the chemicals," the GAO reported in 2006. See GAO-06-1032T, United States Government Accountability Office, *Chemical Regulation: Actions Are Needed to Improve the Effectiveness of EPA's Chemical Review Program*, Testimony Before the Senate Committee on the Enviornment and Public Works, August 6, 2006. See also Nena Baker, *The Body Toxic: How the Hazardous Chemistry of Everyday Things Threatens Our Health and Well-Being* (New York: North Point Press, 2008), pp. 47–48, and Mark Schapiro, "Toxic Inaction: Why Poisonous, Unregulated Chemicals End Up in Our Blood," *Harper's*, October 2007.

215 **Before TRI, no one:** See goldmanprize.org/node/83.

And what a lot they were: Matthew L. Wald, "A Drop in Toxic Emissions and a Rise in PCB Disposal," *New York Times*, March 20, 2009.

216 **It might not surprise you:** U.S. PIRG Education Fund, "Toxic Pollution and Health: An Analysis of Toxic Chemicals Released in Communities Across the United States," March 2007, pp. 4–5.

Ten years ago, the EPA: Yana Kucker and Meghan Purvis, "Body of Evidence: New Science in the Debate Over Toxic Flame Retardants and Our Health," National Association of State PIRGs and the Environment California Research and Policy Center, 2004.

217 **Though critics of REACH:** "Toxic Lobby: How the Chemicals Industry Is Trying to Kill REACH," Greenpeace, May 2006; www.greenpeace.org/toxiclobby.

218 **To give you an idea:** Peter Waldman, "From Ingredient in Cosmetics, Toys, a Safety Concern," *Wall Street Journal*, October 4, 2005. It's not as if the change in Europe came easily. An article in the *Guardian* newspaper warned that REACH would endanger 2 million jobs and send thousands more overseas to less-regulated countries like China. British and other European chemicals firms warned that the proposals were "completely unworkable" and would devastate huge swaths of Europe's chemicals industry. REACH was "going to de-industrialise Europe," said Eggert Voscherau, president of Cefic, the chemical industry's EU lobbying group. "European industry, including the chemicals industry, must not be a test laboratory for a bureaucratic regulatory experiment." Such rhetoric did not sit well with some legislators, who said the chemicals industry was using fear to protect itself from oversight. "Too many in the chemicals industry, and particularly its German lobbying arm, seem to believe that if you are going to tell a lie, then lie big," said Chris Davies, a liberal English member of the European Parliament. "The costs of REACH have been grossly exaggerated from beginning to end. There are still Members here who reject the very idea that industry should bear the burden of proving that the chemicals it puts on the market are safe. There are still Members here who would strip away the testing requirements almost completely: 'Trust us, they are chemical companies' is their argument. There are Members here who still believe that chemicals of high concern should continue to be sold even when safer substitutes are readily available." See www.guardian.co.uk/business/story/0,3604,998303,00.html; www.europarl.org.uk/news/items2005/november17reach.htm; "Toxic Lobby: How the Chemicals Industry Is Trying to Kill REACH," Greenpeace, May 2006; www.greenpeace.org/toxiclobby. **Far from waiting:** Greg Lebedev, "Chemistry Means Business," keynote address for Pittsburgh Chemical Day, May 11, 2004,

reprinted in "Toxic Lobby: How the Chemicals Industry Is Trying to Kill REACH," Greenpeace, May 2006; www.greenpeace.org/toxiclobby.

218 **A Commerce Department memo:** For an excellent summary of the REACH debate, see Mark Schapiro, "Toxic Inaction: Why Poisonous, Unregulated Chemicals End Up in Our Blood," *Harper's*, October 2007.

219 **American public health groups:** From "A Special Case Study: The Chemical Industry, the Bush Administration, and European Efforts to Regulate Chemicals," U.S. House of Representatives Committee on Government Reform—Minority Staff, Special Investigations Division, April 2004; *Chemical and Engineering News*, April 6, 2004.

221 **But California is one place:** Wyatt Andrews, CBS News, May 19, 2008.

222 **In May 2007:** The growing consumer and legislative antagonism to toxic flame retardants has already pressured companies to develop alternative compounds. Some companies, such as the Swedish home furnishing giant IKEA, voluntarily stopped producing products that contain PBDEs ten years ago and are putting pressure on product distributors to start producing alternatives. ("A third of your life is spent sleeping," IKEA's latest brochure reminds you.) Two of the largest manufacturers of brominated flame retardants, Albemarle and Great Lakes Chemical, have moved to develop alternatives to octa and penta. Great Lakes claims that its alternatives will not bioaccumulate. If you'd rather not take a chance on alternative synthetics, a host of companies offer mattresses and couches made with natural latex. See "Albemarle Launches Penta-BDE Alternative," *Chemical Week*, vol. 165, no. 42 (November 19, 2003); "Great Lakes Agrees to Flame Retardant Phaseout," *Chemical Week*, vol. 165, no. 41 (November 12, 2003). For more information on safer electronics, check www.safer-products.org. Toshiba discontinued production of plastics containing PBDEs in its computers and other electronic products. Ericsson has banned deca and other PBDEs from its cell phones and found replacements at comparable costs. Other companies—Apple, Dell, Hewlett-Packard, Panasonic, and Sony among them—are doing the same. What do safer flame retardants cost a consumer? Estimates are that using alternatives would add as little as $4 to a twenty-seven-inch television selling for $300.

223 **The bill was supported:** John Richardson, "Maine to Consider Tracking Toxins in Toys, Products," *Portland Press Herald*, February 27, 2008.

During the hearing, Matt Prindiville: You can read Matt Prindiville's testimony at www.nrcm.org/news_detail.asp?news=2200.

225 **"What's important is that":** *Portland Press Herald*, April 17, 2008. The bill was not as complete as some advocates would have liked. It exempted products made by the pulp and paper industry, for example, as well as food and beverage packaging that is not marketed to children. Of course, compiling all this data—scientific studies of toxics, as well as safer alternatives—will take a great deal of money, and Maine is not a wealthy state. So for now, change will continue to come slowly.

There are other positive signs: Peter Waldman, "Common Industrial Chemicals in Tiny Doses Raise Health Issue," *Wall Street Journal*, July 7, 2005.

226 **None of this can come too soon:** Devra Davis, *The Secret History of the War on Cancer* (New York: Basic Books, 2007), p. xviii.

APPENDIX

229 **When it comes to general rules:** Ellen Sandbeck, *Green Housekeeping* (New York: Scribner, 2007), p. 4.

233 **Be careful when shopping:** Stacey Malkan, *Not Just a Pretty Face: The Ugly Side of the Beauty Industry* (Gabriola Island, B.C.: New Society Publishers, 2007), pp. 119–20.

Select Bibliography

Armitage, Allan M. *Armitage's Native Plants for North American Gardens.* Portland, Oreg.: Timber Press, 2006.

Baker, Nena. *The Body Toxic: How the Hazardous Chemistry of Everyday Things Threatens Our Health and Well-Being.* New York: North Point Press, 2008.

Blanc, Paul. *How Everyday Products Make People Sick.* Berkeley: University of California Press, 2007.

Bond, Annie B. *Home Enlightenment: Practical, Earth-Friendly Advice for Creating a Nurturing, Healthy and Toxin-Free Home and Lifestyle.* Emmaus, Penn.: Rodale Press, 2008.

Bongiorno, Lori. *Green, Greener, Greenest: A Practical Guide to Making Eco-Smart Choices a Part of Your Life.* New York: Perigee, 2008.

Brower, Michael, and Warren Leon. *The Consumer's Guide to Effective Environmental Choices.* New York: Three Rivers Press, 1999.

Brown, Phil. *Toxic Exposures: Contested Illnesses and the Environmental Health Movement.* New York: Columbia University Press, 2007.

Carson, Rachel. *Silent Spring.* Boston: Houghton Mifflin, 1962.

Colburn, Theo. Dianne Dumanoski, and John Peterson Myers. *Our Stolen Future.* New York: Plume Books, 1996.

Commoner, Barry. *The Closing Circle: Man, Nature and Technology.* New York: Bantam Books, 1971.

Daniel, Pete. *Toxic Drift: Pesticides and Health in the Post–World War II South.* Baton Rouge: Lousiana State University Press, 2005.

Davis, Devra. *The Secret History of the War on Cancer.* New York: Basic Books, 2007.

Dolan, Deirdre, and Alexandra Zissu. *The Complete Organic Pregnancy.* New York: HarperCollins, 2006.

Erickson, Kim. *Drop-Dead Gorgeous: Protecting Yourself from the Hidden Dangers of Cosmetics.* New York: McGraw-Hill, 2002.

Fenichell, Stephen. *Plastic: The Making of a Synthetic Century.* New York: HarperBusiness, 1996.

Fitzgerald, Randall. *The Hundred-Year Lie: How to Protect Yourself from the Chemicals That Are Destroying Your Health.* New York: Plume, 2007.

Gibson, Pamela Reed. *Multiple Chemical Sensitivity: A Survival Guide.* Oakland, Calif.: New Harbinger Publications, 2000.

Grossman, Elizabeth. *High Tech Trash: Digital Devices, Hidden Toxics, and Human Health.* Washington, D.C.: Island Press, 2006.

Gussow, Joan Dye. *This Organic Life: Confessions of a Suburban Homesteader.* New York: Chelsea Green, 2001.

Hawken, Paul. *Blessed Unrest: How the Largest Movement in the World Came into Being, and Why No One Saw It Coming.* New York: Viking, 2007.

Hethorn, Janet, and Connie Ulasewicz. *Sustainable Fashion: Why Now?* New York: Fairchild Books, 2008.

Hollender, Jeffrey. *How to Make the World a Better Place.* New York: William Morrow, 1990.

Horton, Richard. "Cancer: Malignant Maneuvers." *New York Review of Books,* vol. 55, no. 3 (March 6, 2008).

Johnston, David R., and Kim Master. *Green Remodeling: Changing the World One Room at a Time.* Gabriola Island, B.C.: New Society Publishers, 2004.

Jones, Samuel B., Jr. *Gardening with Native Wild Flowers.* Portland, Oreg.: Timber Press, 1997.

Kimbrell, Andrew, ed. *Fatal Harvest: The Tragedy of Industrial Agriculture.* Sausalito: Foundation for Deep Ecology, 2002.

Landrigan, Philip J., Herbert L. Needleman, and Mary M. Landrigan. *Raising Healthy Children in a Toxic World: 101 Smart Solutions for Every Family.* Emmaus, Penn.: Rodale Press, 2002.

Lawson, Lynn. *Staying Well in a Toxic World*. Chicago: Noble Press, 1993.

Lear, Linda. *Rachel Carson: Witness for Nature*. New York: Henry Holt, 1997.

Leopold, Donald. *Native Plants of the Northeast*. Portland, Oreg.: Timber Press, 2004.

Louv, Richard. *Last Child in the Woods*. Chapel Hill: Algonquin Books, 2005.

Malkan, Stacey. *Not Just a Pretty Face: The Ugly Side of the Beauty Industry*. Gabriola Island, B.C.: New Society Publishers, 2007.

McDonough, William, and Michael Braungart. *Cradle to Cradle: Remaking the Way We Make Things*. New York: North Point Press, 2002.

McKay, Kim, and Jenny Bonnin. *True Green: 100 Everyday Ways You Can Contribute to a Healthier Planet*. Washington, D.C.: National Geographic Society, 2007.

McKibben, Bill. *Deep Economy*. New York: Times Books, 2007.

———.*The End of Nature*. New York: Random House, 1989.

Michaels, David. *Doubt Is Their Product: How Industry's Assault on Science Threatens Your Health*. Oxford: Oxford University Press, 2008.

National Geographic Society. *The Green Guide: The Complete Reference for Consuming Wisely*. Washington, D.C.: The National Geographic Society, 2008.

Nayak, P. Ranganath, and John M. Ketteringham. *Breakthroughs!* New York: Rawson Associates, 1986.

Pennybacker, Mindy, and Aisha Ikramuddin. *Mothers and Others for a Livable Planet Guide to Natural Baby Care: Nontoxic and Environmentally Friendly Ways to Take Care of Your New Child*. New York: John Wiley and Sons, 1999.

Pollan, Michael. *The Omnivore's Dilemma*. New York: Penguin Press, 2006.

Robson, Kathleen A., Alice Richter, and Marianne Filbert. *The Encyclopedia of Northwest Native Plants for Gardens and Landscapes*. Portland, Oreg.: Timber Press, 2008.

Royte, Elizabeth. *Bottlemania: How Water Went on Sale and Why We Bought It*. New York: Bloomsbury, 2008.

———. *Garbage Land: On the Secret Trail of Trash*. New York: Back Bay Books, 2005.

Sandbeck, Ellen. *Green Housekeeping*. New York: Scribner, 2007.

Schapiro, Mark. *Exposed: The Toxic Chemistry of Everyday Products and What's at Stake for American Power.* New York: Chelsea Green, 2007.

Shabecoff, Philip, and Alice Shabecoff. *Poisoned Profits: The Toxic Assault on Our Children.* New York: Random House, 2008.

Shulman, Seth. *Undermining Science: Suppression and Distortion in the Bush Administration.* Berkeley: University of California Press, 2006.

Steffen, Alex, ed. *World Changing: A User's Guide for the 21st Century.* New York: Abrams, 2008.

Steinberg, Ted. *American Green: The Obsessive Quest for the Perfect Lawn.* New York: Norton, 2006.

Steingraber, Sandra. *Living Downstream: An Ecologist Looks at Cancer and the Environment.* Reading, Mass.: Addison-Wesley, 1997.

Steinman, David, and Samuel S. Epstein. *The Safe Shopper's Bible: A Consumer's Guide to Nontoxic Household Products, Cosmetics, and Food.* New York: Macmillan, 1995.

Sternberg, Guy. *Native Trees for North American Landscapes.* Portland, Oreg.: Timber Press, 2004.

Tallamy, Douglas. *Bringing Nature Home: How Native Plants Sustain Wildlife in Our Gardens.* Portland, Oreg.: Timber Press, 2007.

Tukey, Paul. *The Organic Lawn Care Manual: A Natural, Low-Maintenance System for a Beautiful, Safe Lawn.* North Adams, Mass.: Storey Publishing, 2007.

Weisman, Alan. *The World Without Us.* New York: St. Martin's Press, 2007.

Index

ABOUT THE AUTHOR

MCKAY JENKINS is the author of *The White Death*, *The Last Ridge*, and *Bloody Falls of the Coppermine*, among other books. The Cornelius A. Tilghman Professor of English and Director of Journalism at the University of Delaware, he lives in Baltimore with his family.

ABOUT THE TYPE

This book was set in Berling. Designed in 1951 by Karl Erik Forsberg for the Typefoundry Berlingska Stilgjuteri AB in Lund, Sweden, it was released the same year in foundry type by H. Berthold AG. A classic old-face design, its generous proportions and inclined serifs make it highly legible.